APOLLO MOON MISSIONS

THE UNSUNG HEROES

Billy Watkins

Foreword by Fred Haise

PRAEGER

Westport, Connecticut
London

Library of Congress Cataloging-in-Publication Data

Watkins, Billy W., 1953–
Apollo moon missions : the unsung heroes / Billy Watkins ; foreword by Fred Haise.
p. cm.
Includes bibliographical references and index.
ISBN 0–275–98702–7 (alk. paper)
1. Project Apollo (U.S.)—History. 2. Aeronautical engineers—United States—Biography. I. Title.
TL789.8.U6A6645 2006
629.4′092′2—dc22 2005022473

British Library Cataloguing in Publication Data is available.

Library of Congress Catalog Card Number: 2005022473
ISBN: 0–275–98702–7

First published in 2006

Praeger Publishers, 88 Post Road West, Westport, CT 06881
An imprint of Greenwood Publishing Group, Inc.
www.praeger.com

Printed in the United States of America

The paper used in this book complies with the Permanent Paper Standard issued by the National Information Standards Organization (Z39.48–1984).

10 9 8 7 6 5 4 3 2 1

Contents

Foreword

I have spoken to many different audiences over the past several decades about my experiences as a test pilot and astronaut. A centerpiece of my commentary is on the Apollo 13 mission, which received great notoriety through the Hollywood movie of that dramatic rescue. I find it a great subject in my public appearances to highlight the ingredients that make for success: having the right people and the right training, and working together as a team with the right leadership. I make it a point to ask if anyone in the audience knows the number of people who worked at the peak of the Apollo moon program. The answers I receive are always numbers fewer than 100,000. It is apparent that not many people really appreciate the true size of the team and brain trust that enabled us to travel to the moon. The workforce peaked at over 400,000 people a year or so before we achieved the first landing on the moon during the Apollo 11 mission. Of course, I have noted that a large number of the people I speak to today either were not yet born at the time of the Apollo program or were too young to remember it. The media attention has focused on the astronauts who fly and on mission control personnel who are the visible "real time" participants during the missions. Behind the scenes, solutions to many of the problems that arose during missions leaned on talent from NASA centers and their contractors across the country. Such was clearly the case during my Apollo 13 flight, which offered the team across the country a suite of challenges to be worked out.

But years before the final act—that is, the launch—the team was at work on the design, the manufacturing, the ground testing, and finally the launch prep at Kennedy Space Center. Through this process the team changed somewhat in skill mix and in number, but every step was essential to get to that launch and to achieve mission success. I, along with other astronauts, made numerous visits to contractor plants and facilities to meet personally with the workers on the floor, the people at their "work stations." We took the time to impress upon them the importance of their efforts to get it right. We made sure they knew how the component they were building fit into the overall success of the mission and, in particular, the serious ramifications if it failed. It was clear to me that the workers were awed and dedicated, as I was, to ensuring America's success in carrying out the mandate of our president, John F. Kennedy, to land on the moon. When people ask me about the Apollo 13 mission, their first question is often, "Were you scared?" or, "Did you think you would die?" My answer is "no," an answer that many people have difficulty believing because they cannot conceive of the confidence I had in the capability and dedication of the team. The often-used quip about "flying on a machine built by the lowest bidder" clearly was not the case with the highly directed motivation of the workers during the Apollo program.

My path to the space program was in a way accidental. Through high school and the first two years of college, I was headed toward a career as a reporter. I became enamored with journalism through my assignment as sports editor on my high school newspaper, and I subsequently worked as sports editor and then editor of the junior college newspaper. As with many of us in life, one decision changes everything thereafter. With the Korean War ongoing, I decided it was time to serve my country in that endeavor. This path took me through navy flight training and a commission in the U.S. Marine Corps with an assignment as a fighter pilot. Until that time I had never been in an aircraft of any type and had never had an interest in aviation growing up. Of course, when I grew up, no one could desire to be an astronaut because there were none; there was no space program. But from the first time I was airborne in a navy SNJ training aircraft, I knew this was to be my life.

The strongest emotion I felt, both during the Apollo 13 mission and for a short time thereafter, was disappointment at not carrying out the mission that was planned—a landing in the highlands area of Fra Mauro on the moon. But through a number of public appearances following the flight, including an overseas tour, I became aware of what went on here on Earth during the mission. I spoke to many people who expressed their concern, their hope for our return through prayer vigils in so many places. This realization changed my feeling to one of gratitude and thanks that I had made it back. And as the years have gone by, I have noted that not many people have gone back to the moon. Today I just feel lucky and privileged to have been born at the right time, to have acquired the right background experience, and to have been chosen as a member of the Apollo program—the greatest engineering and exploration program of the twentieth century.

I'm pleased that Billy Watkins, whom I have known for a number of years as a fellow Mississippian, has written this book to highlight some members of our Apollo team who weren't on the media front line. I told Billy that it would be a difficult assignment to choose those to be included in this book, for the Apollo program enjoyed such great success from the talent and contributions of so many hard-working, dedicated people. But Billy's selections were excellent. I learned things about the Apollo program that even I didn't know. The people profiled capture the sense of team spirit, a desire to assure success, and really bring out the human interest side of the program.

FRED HAISE
Apollo 13 astronaut

Preface

This book was born in May 1999 after a phone call from an administrative assistant at the Johnson Space Center in Houston, Texas. I wish I remembered her name. She was confirming my interview the next month with John Young, who commanded Apollo 16 and walked on the moon in April 1972.

I excitedly jotted down the information and thanked her for calling. Before I could hang up, the woman said, "I should tell you something. Captain Young doesn't do interviews anymore. But he pulled your letter out of a huge stack of requests and said, 'Set this one up.' I thought you'd like to know that."

I thanked her again. Then I found a copy of the letter I had written Young about a month earlier. It was a simple, two-paragraph note, asking for an in-person interview for a package I was working on about the thirtieth anniversary of the first moon landing. Cynthia Wall, my features editor at the *Clarion-Ledger* newspaper in Jackson, Mississippi, at the time, had given me the okay to travel to Houston and write the story if Young agreed to the interview. I thank her for that.

And to this day, I still don't know why Young chose my letter over the others. I thank him, too.

Young was my target for a couple of reasons: Neil Armstrong and Buzz Aldrin, the first humans to set foot on the moon, had quit granting one-on-one interviews, and Young is perhaps the most decorated astronaut in history. He flew two Gemini missions. He was the first to orbit the moon alone, on Apollo 10. He spent two days, twenty-three hours, and two minutes on the lunar surface during Apollo 16. He commanded the first shuttle flight, and another after that. And at age sixty-eight, he was the only Apollo astronaut who was still flight eligible.

So around 9:45 a.m. on June 11, I strolled into the public affairs office at the Johnson Space Center and advised the young man behind the front desk that I was there for my 10 o'clock interview with John Young.

"Sir, did no one contact you?" he asked.

My stomach did a free-fall.

"Nobody called me," I said.

"Well, sir, Capt. Young won't be able to see you." (There was a long pause, real or imagined.) ". . . until 10:30. He's in the simulator this morning."

Finally, when my breath returned, I managed a smile and said, "I've waited twenty-four years in the newspaper business to meet a man who walked on the moon. I think I can hold out an extra half hour."

A few minutes before 10:30, I spotted John Young coming up the sidewalk toward the Lyndon B. Johnson Conference Room in Building 2-S. He was wearing a gray suit and moving with the same shoulder-slumped gait I had watched him use to explore the moon's Descartes Highlands, some twenty-seven years earlier. He looked fit and younger than his age.

He was excited about his morning in the shuttle simulator. "Visual systems weren't working, but I made the landing anyway," he said with a smile. "Anytime you can make a landing on the needles [instruments] without visuals, then you haven't forgotten how to do it."

We sat across from each other at a small table. In a corner of the room, inside a glass case, stood the U.S. flag that had flown over the White House the night Armstrong and Aldrin walked on the moon. Young patiently described his memories of exploring another world. He told me about picking up rocks 4.6 billion years old and having moon dust settle beneath his fingernails after each of his three excursions. He and fellow moonwalker Charlie Duke always felt rushed and behind schedule and never really had the chance to stand still and ponder that they were *on the moon!* Time was too precious. "We wanted to go here and couldn't quite make it to there," he said. "I always felt like we were behind in our timeline." The overwhelming silence of the place was one of the things he remembered most. "The only sounds you hear are your [spacesuit's] pumps and fans running," he said. "And if you don't hear them, you know you're in a world of hurt."

Young was thrilled when the flight rotation schedule made him the ninth human to walk on the moon. His number in line perfectly suited Young's longing for anonymity—even in his South Houston neighborhood. "My wife can go in the grocery store, and they just say, 'Suzie, okay,' and take her check," he said, smiling. "I go into the grocery store, and I have to produce my driver's license and my credit cards. And that's okay. I didn't get into this thing to be famous. I got into it to do my job."

I asked him what he thinks about when he looks at the moon now.

His jaws clenched. "Every time I look at the moon, I just can't believe we don't have a base up there right this minute, [and] people exploring and learning new things about our solar system. It's unbelievable," he answered. "If we're really going to explore our solar system, the moon is where we've got to start. Make all the mistakes there. You can get to anywhere in the solar system from the moon, with one-twentieth the energy it takes to get there from earth because of the low-gravity field. It just makes sense to go back and study it, build a base up there, then set our sights on Mars.

"Exploration of space has shown us that we've got a new endangered species out there—and that's *us*." Sooner or later, he warned, the earth would be the target of an asteroid large enough to either destroy it or inflict major damage. "The moon is a good place to put an early warning system for tracking asteroids and comets," he said. "You've got this really cold atmosphere up there at night, and you can look out there and see them without using a lot of fancy [equipment] because that's what you're tracking when you're tracking asteroids—big, cold rocks moving fast. And knowing where they are accurately is the key to saying whether they're going to hit earth or not. That's just one example."

He said we should be figuring ways to move the human race to other places in the universe. "If you've got a base on the moon . . . people in cities and towns all over the moon, you know, [the asteroid] hasn't hurt them," he said. "That's a pretty darn cold-hearted way of looking at it, but I hope it would be my grandbabies and your grandbabies up there."

We talked a while longer, about eighty minutes in all. Near the end of the interview, Young tapped his finger on the table as if an idea had just come to him.

"You know, the astronauts have been interviewed to death," he said. "But there were so many people who helped us go to the moon and bring us back, and nobody knows anything about them. You should find some of those people and write about them. They really did some incredible things."

It took a few weeks for Young's comment to sink in, but the more I thought about it, the more I saw it had book potential. But where would I start?

I had read Andrew Chaikin's masterpiece, *A Man on the Moon*, the story of Apollo told through the eyes of the astronauts. In fact, I found myself reading it again and again. A few weeks after my interview with Young, I noticed in one of the chapters the name of a young flight controller who made a critical decision in the final moments of Apollo 11's descent to the moon. It was Steve Bales. There were no details about him—Chaikin had his hands full just telling the astronauts' stories. But I realized Bales was a perfect example of the unsung heroes Young had told me about.

With the help of some kind-hearted people at the Johnson Space Center, I got Bales' phone number and called him. I think about this a lot. If Bales had been rude or indifferent, I might have taken that as a sign the book wasn't meant to be. But he couldn't have been nicer. Bales was living in New Jersey, totally removed from the space program. We talked several times. The project snowballed from there.

One unsung hero led me to another, then another. And each person I approached seemed genuinely grateful that after three decades someone cared about his or her contributions to the Apollo program. Not a single person turned down my request for an interview. Not one. Some said it was the first time they had ever been asked, which I found astounding.

I had a difficult time deciding on the number of unsung heroes to spotlight. After all, hundreds of thousands deserved to have their stories told. I finally settled on fourteen—one each for the twelve men who walked on the moon, and two more for Jim Lovell and Fred Haise, who spent years preparing to do so on Apollo 13 only to see their dreams ruined by an explosion while coasting to the moon. I hope all those who worked behind the scenes to help make Apollo successful will somehow see themselves in these stories.

The interviews took place between late 1999 and early 2005. At least two people featured in this book, Julian Scheer and Joseph Laitin, passed away before it was published, and I am deeply saddened they never got to see the finished product. I hope their families enjoy it and pass it down for generations to come.

Manned spaceflight was not a foreign subject for me to tackle. I was a space fan growing up. My mother, God bless her, let me skip school to watch the first few launches. I created my own Mercury and Gemini cockpits, which consisted of a recliner, a Styrofoam space helmet, a window view created from pictures of the earth taken by astronauts and published in *Life* magazine, and an instrument panel made of a large square piece of cardboard. I had drawn the array of gauges and buttons from a picture I saw in a book, and my older brother, W. G., came up with the ingenious idea of using toothpicks as switches. I still have the instrument panel.

By the time Apollo rolled around, having reached my teens, I retired from the cockpit recliner. But I passionately watched every mission. Apollo 8, in December 1968, remains especially memorable for me. My mother was in the hospital on Christmas Eve that year, and my stepfather was spending the night with her. W. G. had married and moved away. I was fifteen years old and, except for human voices coming from a tiny spacecraft in orbit around the moon, I was alone at my house in rural Gholson, Mississippi.

When Frank Borman, Jim Lovell, and Bill Anders read aloud the first ten verses of Genesis as I stared at the fuzzy television pictures of the moon that night, a peace enveloped me. After the telecast, I grabbed the Bible off our coffee table and read the verses again and again. The broadcast from lunar orbit deeply touched me. What had started out as one of the worst Christmas Eves of my childhood turned into the most special.

Never did I dream that I would one day write the story of how the reading of Genesis came about, or that I would have the chance to tell Jim Lovell and Bill Anders how much their message that night in 1968 comforted a lonely young boy back on earth.

Those are just two examples of my amazing journey while researching and writing this book. All because John Young, who retired from NASA on the final day of 2004, decided for whatever reason to grant me an interview.

BILLY WATKINS
May 2005

Acknowledgments

A book, I have discovered while writing my first, is not simply a person sitting down and pounding keys until it is finished, though there are countless hours of that. It is a product of many people who share a passion. And I have many to thank.

My agent, Jim Dickerson, promised me this day would come. Jim, you're a man of your word, and a treasured friend. Dr. Heather Staines, senior project editor at Greenwood Publishing Group and a true professional, believed in this idea from the start. Heather, you're the best. I would also like to thank others at Greenwood: Andrew Andrusyszyn, Nicole Azze, Marcia Goldstein, and anyone else whose fingerprints are on this book. Thanks to Jane McGraw at Capital City Press.

I am forever indebted to fellow Mississippian and Apollo 13 astronaut Fred Haise, who read every chapter for technical accuracy. His knowledge, patience, and attention to detail were never ending. He would even send e-mails reminding me to "get some sleep." Freddo, thanks for always being there.

Thanks to Margaret Persinger at Kennedy Space Center and Mike Gentry at Johnson Space Center for their help in providing photographs. Also, thanks to Carol Butler at the NASA history office.

I'd like to thank Apollo moon voyagers Jim Lovell, John Young, and Bill Anders, along with flight director Glynn Lunney, for their inspiration and time.

I was fortunate to have many friends who encouraged me throughout the writing of this book and, in ways they may never know, helped make it possible: Larry Conner, Bill Zimmerman, Keith Warren, Amanda

Brice, Rick Cleveland, Danny and Gail Lynn, Chris Todd, Gary Pettus, Grace Shumaker, Marshall Ramsey, John Temple, and Jerry Mitchell. Thanks, also, to my brother, W. G., and sister-in-law Polly. Your words always seemed to find me at the right time.

Of course, I would like to thank the fourteen people profiled in this book for their willingness to share their stories. Also, thanks to Rosemary Roosa—daughter of Apollo 14 command module pilot Stuart Roosa—for her encouragement and help.

Thanks to my mother, Annie Margaret Watkins Hailey. No son has ever felt more loved. Thanks, also, to my stepfather and friend, Levin "Bubba" Hailey.

To my children—Mandy, Todd, and Taylor—and my son-in-law, Kevin: thanks for your love and support.

To my dad, Jimmy: I wish you had lived to read this.

And most of all, thank you to my wife Susan, who made sure the house was quiet when I was trying to write, read every chapter and offered suggestions, and *always* believed in me. This one's for you.

Introduction: History of Apollo

In a poll to identify the most important news events of the twentieth century, Americans voted the U.S. moon landing in 1969 third, behind only the atomic bombing of Japan in 1945 and the Japanese's attack on Pearl Harbor in 1941. Women and journalists ranked the moon landing second; the Wright Brothers' maiden airplane flight in 1903 was first. People under the age of thirty-five voted the moon landing number one, followed by the bombing of Japan, the 1963 assassination of President John F. Kennedy, the Wright Brothers' flight, and Pearl Harbor.

The results indicate those furthest removed in time from the lunar voyages are the most intrigued by them.

"I think you'll continue to see that," says Jim Lovell, an astronaut who twice flew around the moon during America's Apollo space program. "Most events that have a great effect on the people of the earth run in a cycle whereby there is initial euphoria, followed by a period of forgetfulness until it matures.

"It's like a wine, in many ways. A wine master develops this new wine; he's pleased with its taste. But it still has to age before it becomes a *really* good wine. Same way with the Apollo program. It will be for the historians to finally come back and realize what we did."

Glynn Lunney, a Pennsylvania native and a flight director at the National Aeronautics and Space Administration's (NASA's) mission control during Apollo, agrees: "I'm actually surprised [the landing] rated as high as it did. As we get further away from Apollo, it will look even more like the top story of the twentieth century. Wars come and go. History books are one war after another. This Apollo thing was unique.

"I believe in years to come, school kids will come to know the date 1969—the first moon landing—the same way they know 1492 when Columbus discovered America. It will prove to be that important."

This was Apollo: Eleven flights of three-man crews between October 1968 and December 1972. Six moon landings. More than 850 pounds of lunar samples gathered by the twelve astronauts who flew to the surface. Humans' first views of earth as a sphere. No Americans lost in space.

It was the end result of a dream and a challenge issued to Congress by President Kennedy in May 1961 of landing a man on the moon by the end of the decade and returning him safely to earth.

Lunney, an engineer, and his cohorts at NASA were stunned by the president's lofty goal. They were in the beginning stages of the Mercury program, trying to figure out how to get one astronaut, strapped inside a tiny capsule, into orbit. So far, they had managed to send Alan Shepard into space on a fifteen-minute, suborbital mission.

"The whole moon goal was staggering," Lunney recalls. "Here we were struggling to get a 2,500-pound capsule up, and this thing he just assigned us was going to require getting 250,000 pounds into earth orbit. And in our business, weight is a hell of a big factor. Looking back on it, it's just amazing that people [in power] were able to agree, 'We're going to the moon before the decade ends.'"

But the Soviet Union, a communist country, was already making quiet claims on the territory of space. In September 1957, the unmanned satellite Sputnik was blasted into earth orbit. In April 1961, twenty-seven-year-old cosmonaut Yuri Gagarin became the first human to enter space and circle the globe aboard Vostok 1. The missions were conducted in secrecy and publicized only after success was assured. This added to the mystique. And if the Soviets could conquer the skies, many Americans reasoned, certainly they would eventually rule the world.

In a grand strategic move, Kennedy raised the ante just four months after taking office. "He redefined the rules of the space race by challenging the guys who got into space first by saying, 'Here, try to knock this chip off my shoulder. We're gonna go to the moon, land, and return within the decade,'" Lunney says. "When you think about it, it was such a staggeringly bold decision and pronouncement that it made the United States and the Soviets both start from scratch again. Neither of us knew how to go about going to the moon, so in a way, nobody was ahead in the space race.

"President Kennedy obviously had an inherently large faith in America and what we could do when we put our minds to it. Plus, I think the president wanted to take a step that was visible, clear, and compelling to people. And it was strange in that once a time frame like that had been set and embraced nationally, the politicians lost control of the money. Engineers now decided how much money we needed to reach our goal. It was a unique time."

This was Apollo: A classic lesson in the old adage that it's not how one starts that counts, but how one finishes. From July 1961 through May 1963, the United States flew five manned Mercury missions, four of them orbital. Gordon Cooper circled the earth for thirty-four hours on the program's final flight.

"The Mercury capsule was kind of a '50s machine," Lunney says, "and we probably did all we could with that vehicle. It didn't have much in the way of propulsion. It had very little attitude control system; not much in the way of digital equipment onboard."

Mercury was a basic course in spaceflight for NASA.

The Soviets flew only three missions during that period, but two of them were in space at the same time and performed the first rendezvous by maneuvering within five miles of one another—a critical procedure in going to the moon. And while NASA took twenty-two months to prepare for the two-man Gemini flights, the Soviets flew four more missions. They launched the first woman into space, flew the first three-man crew aboard the more elaborate Voskhod spacecraft, and cosmonaut Alexei Leonov performed the first spacewalk on Voskhod 2 in March 1965.

But something wasn't right with the Voskhod spacecraft. All three cosmonauts suffered from space sickness on Voskhod 1, and the Voskhod 2 crew of Leonov and Pavel Belyavyev landed so far off course, it took two days for them to be recovered in the Ural Mountains.

While the Soviets scratched their heads, NASA's Gemini program took off—ten missions between March 1965 and November 1966. U.S. astronauts performed spacewalks, learned the intricate procedures of rendezvous and docking, and set an endurance record of thirteen days on Gemini 7.

"Gemini was similar to Mercury, but it was a big step in a lot of ways," says Lunney, a flight director on the final four Gemini missions. "We had a lot of digital equipment onboard—radar, a digital computer for navigation and guidance. We had fuel cells [that produced electricity] and docking systems [to link with other vehicles].

"It may have all looked smooth to the public, but we were doing things every way from Sunday. We had innumerable problems. Thrusters would clog all the time; fuel cells used to flood with water. We did rendezvous where we used large amounts of fuel and didn't know why. We didn't know how to do EVAs. . . . [Gene] Cernan almost didn't make it back into the spacecraft on Gemini 9 he had so many problems. He was working so hard to maintain his body position because we had not imagined the need for handholds or footholds. We just sort of imagined people floating to wherever they needed to go, and it didn't work like that. Buzz [Aldrin] had handholds and footholds on Gemini 12, and they worked beautifully.

"Let me tell you, a lot of people think Gemini was some sort of 'filler' program between Mercury and Apollo, but that's just not the case," Lunney continues. "Without Gemini there would have been no Apollo. When we flew, ground controllers were grappling with all kinds of systems problems, different kinds of rendezvous, and all of it just kept expanding our experience base. On Gemini 12, the radar failed before we got into the rendezvous, so we flew without it. When you start getting a lot of stuff like that under your belt, you start to realize your backup techniques, how the crew handles it, how the guys on the ground handle it. We wound up with some pretty savvy people. By the time we came out of Gemini, we were a very tough, confident group of people."

This was Apollo: A brutal reminder that trying to go to the moon is dangerous business.

On January 27, 1967, while performing a simulated launch of Apollo 1 at Cape Canaveral, Florida, astronauts Gus Grissom, Ed White, and Roger Chaffee died in a cockpit fire. A spark ignited nylon netting beneath Grissom's left couch, and flames quickly engulfed the spacecraft's pure oxygen environment. The crew perished within seconds. Investigations and congressional hearings soon followed. Engineers and astronauts, determined to carry on, suggested more than 1,300 changes in the Apollo command module to make it safer. It would take months to redesign the spacecraft.

The race to the moon now seemed to favor the Soviets, who had developed their own new three-seat spacecraft known as Soyuz. In April 1967, Vladimir Komarov was launched into earth orbit to test its systems, and hardly anything worked correctly. Komarov had trouble communicating and maneuvering the craft. He died when the main parachute failed to deploy during re-entry.

Both sides were now trying to deal with personal tragedy and new technical challenges.

This was Apollo: More than 400,000 people, from California to Florida, working with focus and purpose rarely seen in world history.

"The '60s, for the most part, were a terrible time in America," Lunney says. "President Kennedy was assassinated. Martin Luther King, and [Robert] Kennedy were assassinated. I still think there was a significant fear of the Soviet Union through most of the '60s. We were caught up in the Vietnam War. There was the Democratic Convention in Chicago, the protesters outside, and footage of police officers just whacking the hell out of 'em. We had the hippy movement going on, the free-love thing, the drug stuff.

"But for those of us involved in getting to the moon, it was like we were on this little island doing our thing. We could look out and see what was happening on the mainland. We saw the events on TV, and we were affected by them. But not as much as most people. It's shameful to say that, but it's true. We were so consumed by getting to the moon, we had little time to pay attention to anything else."

This was Apollo: A vehicle that stood thirty-six stories high on the launch pad and weighed 3,000 tons. At its base was the most powerful rocket in history, the three-stage Saturn V developed by Wernher von Braun. The Saturn V was the single biggest factor in America winning the race to the moon. Its first stage produced 7.5 million pounds of thrust, or the equivalent of 543 fighter jets. Once the astronauts were in earth orbit, the rocket's third stage was capable of producing the magic speed of 25,200 miles per hour to escape earth's gravitational forces and send astronauts on a coast toward the moon. The Saturn V would fly thirteen times—two test flights and ten Apollo missions (Apollo 7 used the smaller Saturn IVB)—without a failure.

The Soviets tried to answer with the powerful N1 rocket, but unmanned test firings, in February 1968 and early in July 1969, ended with the vehicle plummeting back to earth only seconds after liftoff.

Seventeen days after the second N1 catastrophe, the fifth Apollo mission—Apollo 11—landed on the moon's Sea of Tranquility.

America had won.

"And to that small fragment of people who say the moon landings were a hoax, I offer two words of defense—Soviet Union," Lunney says. "Don't you believe that if America hadn't landed on the moon, the Soviets would have been the first ones to cry out about it? It's not like they haven't tried to disprove it. They *know* we did it."

This was Apollo: A cone-shaped command module, which served as the living quarters for three astronauts and was the only part of the launch vehicle that returned to earth. It was an incredible spacecraft, especially after the redesign, capable of precision maneuvers in space, withstanding the 5,000-degree (Fahrenheit) heat of re-entry and floating in the Pacific Ocean after landing.

One of the things it couldn't do was land on the moon. That was left to the lunar module (LM), one of the most unique machines in aviation history. It flew only twice before the first landing. Stored in the top of the Saturn's third stage and captured by the command module shortly after leaving earth orbit, the LM resembled a flying spider with its four landing legs and two windows that looked like eyes. It had both a descent and ascent engine and was the only space vehicle ever flown by astronauts standing up. Seats were removed because of weight-critical issues.

After the command module and lunar module entered lunar orbit, two astronauts would transfer over to the LM and fly it to the surface. One astronaut would remain behind in the command module. Once the explorations were completed, the command module and lunar module would meet again. Through a connecting tunnel, the moonwalkers would transfer the samples they had gathered and re-enter the command module. The lunar module was then jettisoned.

This was Apollo: On the night of July 20, 1969, inside a bulky white spacesuit, a thirty-eight-year-old native of Wapakoneta, Ohio, descended a nine-rung ladder toward the moon. The man stood five feet, eleven inches and weighed 165 pounds. He had blonde hair and blue eyes. He was a husband and father of two. He had flown seventy-eight combat missions in Korea as a Naval aviator. His hobbies included playing the ukulele, and his $30,054 annual salary as a civil servant was nearly the combined yearly income of his Air Force crewmates—Colonel Edwin E. "Buzz" Aldrin, who flew to the surface with him, and command module pilot Lt. Colonel Michael Collins.

"That's one small step for man," Neil A. Armstrong said, as he placed his left boot on the moon. "One giant leap for mankind."

Three months later, the Soviets flew three Soyuz spacecraft into orbit around the earth at the same time, with seven cosmonauts aboard. The biggest news from the mission: Valery Kubasov performed the first welding experiment in space.

PART I

"The *Eagle* Has Landed"

1 STEVE BALES

Guidance Officer, Apollo 11

Steve Bales knew Neil Armstrong was a proven test pilot and astronaut. He was positive NASA had done its homework before selecting Armstrong to command the first lunar-landing mission and to take the first steps on the moon.

But there was one thing that Bales, chief guidance officer for the historic Apollo 11 flight, couldn't get out of his mind: Armstrong seemed so . . . old.

"Most of the people in mission control were twenty-five, twenty-six, twenty-seven years old," Bales says. "I was twenty-six. Neil was thirty-eight. That seemed ancient to me."

The manned space program during the race to the moon was a strange mix, indeed. In the limelight were the crusty, veteran astronauts, many of whom had flown combat missions. Behind the scenes were the young, creative engineers who served on the ground support teams. Diversity was created by urgency. Striving to meet President Kennedy's 1961 challenge to land a man on the moon before the end of the decade, NASA had to find the brightest aerospace engineers available, train them quickly, allow them to learn on the go, and—most of all—trust them. Many of the top candidates were just out of college, fresh products of the most challenging and up-to-date engineering curriculums in the country.

"The fact is, there were no experts on the subject [of landing a man on the moon] of any age," Bales says, "and when you're in your twenties you're usually more ready to take on a stressful challenge."

Bales was one of the chosen few, joining NASA in December 1964, just before the second unmanned Gemini mission was launched.

STEVE BALES

Steve Bales was just twenty-six years old when he made a critical call as guidance officer on Apollo 11 to help save America's first attempt to land on the moon. Bales later became deputy director of operations at the Johnson Space Center in Houston. [Courtesy Steve Bales]

And on July 20, 1969, with astronauts Armstrong and Buzz Aldrin speeding toward the lunar surface minutes from arguably the most historical event of the twentieth century, the entire mission's fate fell on the shoulders of Steve Bales.

When a computer guidance alarm sounded, Bales had fifteen seconds to decide whether to continue the mission or call for an abort.

Fifteen seconds.

Twenty-six years old.

The world and two anxious astronauts awaited his call.

Bales grew up in Fremont, Iowa, a town of about 600 residents in the southeastern part of the state. His dad was the school janitor. His mother worked as a beautician.

Bales loved to read space books when he was a child, and at the age of thirteen, he saw a Walt Disney television show about how people might fly to the moon one day. "This show, probably more than anything else, influenced me to study aerospace engineering," he says. "And this wasn't the ordinary thing to do for a boy raised in a small Iowa farming community in the '50s, but it does show how strong and powerful an idea can be, if it falls on fertile ground."

Bales went straight to NASA from Iowa State University's aerospace engineering program. "NASA was trying to get people in college to understand what the space program was all about," he recalls. "So they made a pretty smart decision: They hired maybe 400 to 500 engineering students for summer internships, just to get a look at what was going on.

"I went down there in 1964, and we did all sorts of things—plotting data, writing small reports, something you could give a college student to do. Nothing critical.

"This was right between the Mercury and Gemini programs. They had flown the Mercury flights out of a control center in Florida. It was tiny and totally inadequate to fly Gemini, certainly inadequate to fly Apollo and the lunar flights. So they were frantically trying to build this new control center in Houston, the software for the control center, the engineering labs. It was wild.

"As soon as I got back to school, I wrote them a letter and said I'd like to come to work there some day. I'd thought one time that I wanted to work on jet planes, but when I saw the possibilities at NASA and what they were trying to do, I knew it was for me.

"They were willing to hire young people and give them plenty to do. And it was challenging. It was like, 'We have a tremendous problem—figuring out how to land a man on the moon—and we've got to solve it quickly.' And that's pretty much the way it was around NASA, you either helped figure it out, or you went to work somewhere else."

Despite his boyish looks, Bales quickly became a main player. Three months after arriving at NASA, he was allowed to serve as a backup flight controller for the Gemini 3 and Gemini 4 manned missions. Twenty months after being hired, he was in the main seat as guidance officer for Gemini 11 and 12.

Then came the Apollo program and the tragic fire during a launch simulation in January 1967 that killed astronauts Gus Grissom, Ed White, and Roger Chaffee. The program was put on hold while NASA sorted out how it was going to recover. Bales used the downtime to study the two new Apollo spacecraft—the command module, which would carry three astronauts into lunar orbit and back, and the lunar module, which would be used for the actual landing.

"The guidance systems for Apollo—the computer, navigation platform, navigational telescope, landing radar—were much more complicated than Gemini's," Bales says. "Today, the Apollo computer wouldn't even measure up to the simplest home computer. It had a memory of 64,000 words and performed only about 1,000 instructions per second. But in 1967, it was considered really something in the world of flight systems avionics. The people at MIT who programmed it lavished care over each and every instruction, and we who were going to monitor it in flight had to know what it was going to do at every turn."

Once the Apollo program was up and running again, Bales was the midnight-shift guidance officer on Apollo 7, which made its three-man maiden voyage in October 1968. He was lead guidance officer for Apollo 10, which carried astronauts Tom Stafford and Gene Cernan within 50,000 feet of the moon; engineers were still tweaking the lunar module in preparation for actual landings.

"Apollo 10 did everything but land," Bales says. "So in the control room, it was important that all of us, who were going to work on Apollo 11, also got to work Apollo 10. Everything you did in preparation, except for landing, was the same. The lunar module had to be powered up. The computer had to function correctly. The LM had to fly down close to the moon, then rendezvous later with the mother ship. The flight software wasn't prepared for a landing, and the vehicle wasn't structurally ready to land, either. But that flight proved the timeline and plan we had developed would work. And we were lucky to have Tom Stafford as commander of that flight. He was a super, super guy and solid as a rock."

Apollo 10 returned to earth on May 26, just fifty days before Apollo 11 was to blast off, so the training schedule for Apollo 11 was tight and intense, with no room for distractions. President Nixon invited the astronauts to the White House for a sendoff dinner; NASA politely declined. Space officials felt the time would be better spent in training.

Mission control teams were divided into rotating units for different stages of the flight. Flight controller Gene Kranz's team, to which Bales was assigned, would handle the most critical phases—lunar landing and ascent. Bales' teammates were Jay Green, flight dynamics; Don Puddy, telecommunications and environmental LM systems; Bob Carlton, LM propulsion systems; Chuck Berry, surgeon; and astronaut Charlie Duke, capsule communicator. (Duke would eventually walk on the moon as LM pilot of Apollo 16, along with John Young.)

Just as he had done on Apollo 10, Bales' job was to monitor the LM's computer, landing radar, and trajectory. "We trained six days a week, twelve hours a day," Bales says. "The landing team trained two days a week, the launch team one day, then the ascent team one day. You had the prime [astronaut] crew, plus the backup crew, working. Pretty soon we were running out of time."

During the first landing simulation, Bales was busy at his console when he noticed someone out of the corner of his eye sit down next to him. It was Chris Kraft, director of flight operations for the Manned Spaceflight Center.

"The boss," Bales says, "plugged in right beside me. And he did that to everybody on the team, at one time or another. He wanted us to know that he was taking a personal interest in who was sitting at those consoles, and he wanted to know how good we were."

Simulations were done not only to find out how the astronauts would react to sudden problems, but also how the flight controllers handled stressful situations. How fast did they decipher information fed from the spacecraft? Were they able to express themselves to the flight controller and capcom precisely and quickly? Did they over-react? Under-react? Simply, were they good enough?

"Looking back, the sims were almost as pressure-packed as the flights," Bales says. "Sometimes [the simulation supervisors] would throw problems at us that were so hard, things wouldn't turn out so [well]. We'd crash or we'd abort the landing. Then we'd go into this meeting room, and Kranz would go around the room and say, 'Why did you do that?' And everybody would have to explain the logic behind their decisions. Then the supervisors would tell us what we did wrong on that particular problem; that maybe we acted too soon or too late. There was quite a bit of learning and quite a bit of pressure.

"That's the way it went for two months, which I think helped prepare us mentally. You have to remember that nobody had ever landed on the moon before, so we were sort of making up how to do all these things. Everybody was learning together. And we practiced right up until two days before the launch."

Liftoff went without a glitch on July 16. Once the astronauts successfully performed their Translunar Injection burn, setting them on a sixty-six-hour coast to the moon, Bales figured it would be a good time to review procedures, remind himself of critical steps during the lunar descent, and reflect quietly on what was about to happen. Instead, mission control bordered on chaos during those three days.

"Actually, that was probably the hardest time for the flight team," Bales says. "There were 500,000 people working for NASA at that time, and after the spacecraft lifted off, 498,000 didn't have anything to do but sit around and worry. The 2,000 of us in flight operations didn't want to talk to anybody, but somehow the design engineers kept finding a way into the control room. They'd say, 'If this happens, how are you going to handle it?' And we'd stare back at them and say, 'Where have you been for the past six months?'

"They would come in with possible failures and ask if our mission rules had guidelines for them. Sometimes, yes. Sometimes, no. We weren't exactly flying by the seat of our pants, but there was only so much time to work out contingency plans. All the design engineers had been so totally involved in designing and building software and checking things out, they hadn't had a lot of time to pay attention to what we in operations were doing. Now, as time got closer to land, they took a big interest in our procedures."

Bales slept in the bunkroom at the Manned Spaceflight Center on the eve of the landing. He reported for duty at 7 a.m., one hour before his shift was to begin and about eight hours before Armstrong and Aldrin were scheduled to land on the moon. He wore what he calls "the standard NASA uniform"—gray slacks, short-sleeve white shirt, narrow black tie, and a blue sports coat. He removed the jacket as soon as he arrived at mission control.

As hard as Bales tried to treat it like another day at the office, he couldn't. He went over and over his notes and checklists. He smoked one cigarette after another.

"You could've cut the tension in that room with a knife," he says. "The control center had a big viewing area in the back for VIPs, and I recognized most of the people in there. Everybody who was anybody in the aerospace field was in that room, including [rocket guru] Wernher von Braun. There was tons of pressure. I mean, can you imagine how sad a day that would've been if somebody on the ground had made a mistake, or some little piece of equipment failed, and the landing had to be scrubbed?

"So to counteract the pressure, you just concentrated on all the things you had to do. And there was plenty to do." Bales was seated on the end of the front row at mission control, which was commonly known as the trench. Flight dynamics officer (FIDO) Jay Green sat to his immediate right.

At 12:46 p.m., on July 20, Armstrong and Aldrin undocked the LM nicknamed *Eagle* from the moon-orbiting command module *Columbia,* piloted by Michael Collins. As the two spacecraft disappeared around the far side of the moon and out of contact with earth, Kranz asked that all

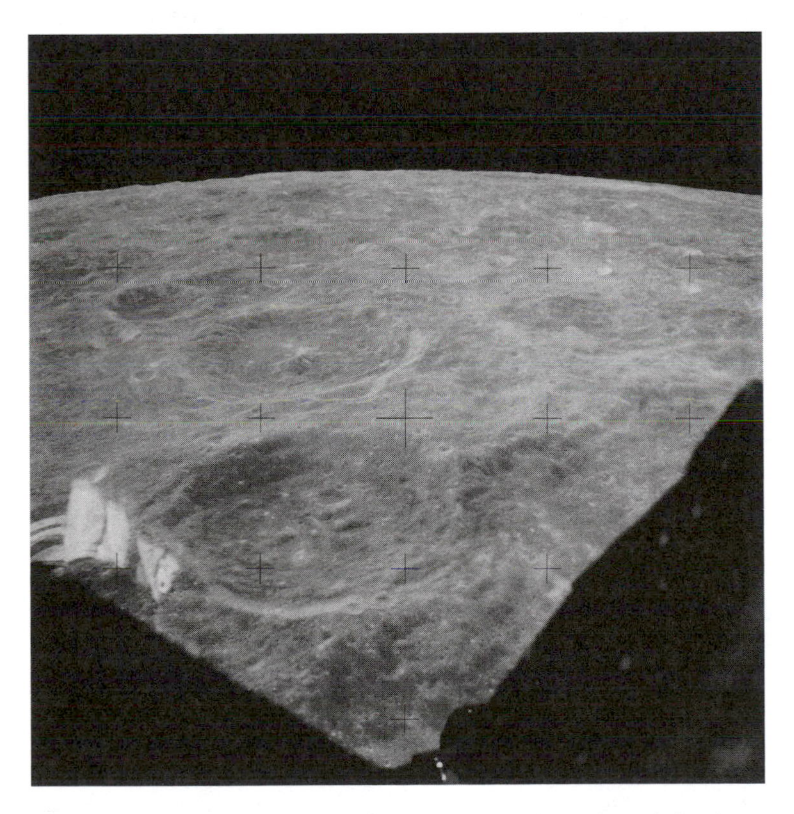

The moon as seen through Buzz Aldrin's window aboard the lunar module *Eagle* during powered descent. Commander Neil Armstrong had to fly the LM over numerous boulder fields before finding a smooth landing site. [Courtesy NASA]

controllers switch to his private loop. He gave a short speech to his young team: "You are well trained, and we are here to do something that no one else has ever done. I am totally confident in you. But no matter how it turns out today, I want you to know that I am totally behind each of you when we walk out of this room."

On the moon's backside, *Eagle* began a deceleration burn that took it down within 50,000 feet of the surface. At that point, ground controllers examined all of the spacecraft's systems; all were given a "Go" for landing. Shortly after 3 p.m., the LM's descent engine again was fired, sending Armstrong and Aldrin on a thirteen-minute journey toward the moon's Sea of Tranquility.

Immediately, there was trouble.

"We started the vehicle off in the wrong place on the initial burn, and we also started it off at the wrong velocity," Bales says. "It was nobody's fault, really. There was unexpected venting out of the vehicle on the backside of the moon that changed its velocity at the time of engine ignition. Maybe they got a little bigger kick when they undocked than expected. So we wound up going toward the moon thirteen miles per hour faster than we should have been.

"If that figure got up to twenty miles per hour, we'd have to abort because the computer [wouldn't] realize this, and it could fly them right into the ground. We had ground-system radar in real-time, which was a phenomenon in itself—tracking this little thing just a few miles above the moon.

"So we programmed the computer to correct the problem, which it did. At that point, we were in really good shape. I was a bit shell-shocked, but the descent was going great."

As he tried to calm his nerves, it occurred to Bales that he no longer viewed Armstrong as old. Now, he was appreciative of all that experience at the controls of the *Eagle*.

"I never knew him well," Bales says of Armstrong. "When Joe DiMaggio passed away, they said he was an enigmatic man. Well, Neil Armstrong is an enigmatic man too—just a hard person to get to know.

"He's a nice fellow. He was always pleasant and nice to me. And apparently he was fearless. There was lots of evidence of that before he came to NASA. He flew combat missions in the Korean War. He flew the X-15. He was the guy for this job."

Eagle was 33,000 feet above the moon when Armstrong calmly radioed Houston, "Program alarm."

Bales heard only the first word.

"It's a 1202," Armstrong added.

"Just as he said that, I looked up at my console, and there was a big program alarm light and a 1202," Bales recalls. "And I started trying to remember what rule we'd written for that alarm."

More than 600 engineers had helped design and maintain the computer system, so it was understandable that Bales couldn't pinpoint the exact problem in a split second. There were simply too many possibilities. Collins, orbiting directly overhead in *Columbia* and listening in, couldn't remember ever hearing of a 1202. He scrambled to see if it was listed somewhere in the flight plan.

"Give us a reading on the 1202 Program Alarm," said Armstrong, his voice showing a hint of urgency. Armstrong's request was aimed specifically at Bales, who suddenly found the mission—and history—squarely in his lap.

Any alarm at that point was cause for great concern because nobody had ever been where Armstrong and Aldrin were—flying in a newly designed spacecraft and heading for the tall boulders and deep craters of the moon. "We were in a gray area at that moment," Bales says, "and we always knew that a gray area could turn to black in an instant."

Bales suddenly recalled a simulation a couple of weeks earlier when a similar alarm (starting with the number 1) had come up, and he had ordered an abort. "Right after that, we had a hot debriefing," Bales says. "They asked me, 'Why did you abort?' And my answer was, 'Why shouldn't you?' The fact was we didn't have rules for that particular class of alarms. They were put in when the software was first written to help debug the system. After that, we just sort of forgot about them because we thought they'd never come up again." After the simulation, it was written into the flight rules that any alarm starting with the number 1 was considered "important, but non-critical."

Over the headset, Bales heard one of his backroom guidance specialists, Jack Garman, say: "Steve! Steve! It's a 1202. It's the if-it-doesn't-come-up-too-much rule."

"It was exactly what I needed to hear, and I agreed with him," Bales says. "We had said if the alarm didn't come up too often, we'd keep going."

A 1202 meant that the computer guiding the spacecraft was being asked to perform too many tasks. "The computer was on a fixed data cycle, and every second it had certain things to do," Bales explains. "It would ask itself, 'Have I done everything?' If it had, it would go on to the next second. If it failed to get everything done in that second, it would stop what it was doing and shoot out this 1202 alarm.

"Remember, this was 1969, so all the computer had room to do on the display was print out numbers. It didn't print out English. Then you had to know what the numbers meant."

Nine seconds after Armstrong asked for a reading, and about six seconds before an abort became the only option, Bales weighed Garman's recommendation and his own analysis. He radioed capcom Charlie Duke in an excited, squeaky voice, "We're GO on that alarm." Duke immediately relayed the message to Armstrong and Aldrin.

Four minutes later, with *Eagle* 3,000 feet above the moon and descending at twenty-seven feet per second, another computer alarm sounded. This time it was a 1201. Without hesitation, Bales said to Duke, "We're GO on that alarm." Duke hurriedly gave Armstrong and Aldrin the good news: "We're GO. Same type. We're GO."

Three-and-a-half minutes later, at 3:17 p.m. Houston time, Armstrong maneuvered the spacecraft over a treacherous mountain region to a soft touchdown in the Sea of Tranquility. His first words from the surface remain an icon for the space program: "Houston, Tranquility Base here. The *Eagle* has landed."

Ground controllers and the Apollo astronauts who had gathered in mission control were visibly shaken—not only by the fact that Apollo 11 had landed on the moon, but they understood what a masterful piloting feat Armstrong had just pulled off. At the time, no one in NASA admitted to the media how dangerously close the *Eagle* came to running out of fuel. Armstrong, whose heart rate was 150 beats per minute during the final seconds of descent, had trouble finding an area smooth enough to evenly support the four landing pads (they had to be level for the takeoff twenty-one hours later), so he had to fly farther and longer than expected. "Best anyone can figure, Neil had about seven seconds of fuel left," Bales says. "Again, his experience paid off."

Through the years, Bales has replayed his own role in those final, anxious moments and weighed his decisions dozens of times.

"I was 100 percent sure in my heart that I was doing the best I could at the time," Bales says. "Hey, we'd never done an abort, either. So that wasn't a sure thing. The risk of going onto the surface was probably less than the risk of trying to do an abort.

"What we would never do is send the crew down to the surface with a failed computer. But with this alarm, we didn't have enough evidence to know that the computer was absolutely sick. In that situation, you have to balance out all the risks, and in my mind, it was safer to continue. There were risks no matter what we did. All I could do was implement the rules, the very best I knew how, and pray that it turned out OK."

He often asks himself the obvious question: What if they had crashed?

"It certainly would've been the end of me at NASA," he says, candidly. "I would've quit. I wouldn't have wanted to stay around, even if people

Buzz Aldrin salutes the American flag shortly after he and Neil Armstrong planted it in the lunar soil on Apollo 11. [Courtesy NASA]

had been nice about it. I would've testified to some investigation panel, and then I would've left. I would've gotten lost somewhere and found something else to do. But it would've been with me the rest of my life."

What does a twenty-six-year-old engineer, who just played a major role in the first landing of humans on the moon, do with the rest of his life?

Bales wondered that a lot.

"I was part of Apollo 8—not a big part, but I was there and got caught up in the emotion of sending three men to orbit the moon on Christmas Eve," Bales says. "But the emotion and excitement involving Apollo 11 were absolutely unbelievable. People outside the program just can't imagine what being a part of that was like."

Bales worked as guidance officer on Apollo 12, which landed the late Pete Conrad and Alan Bean on the moon in November 1969. He helped out on other Apollo missions, including the near-tragic Apollo 13 that didn't land because of an electrical explosion on the way to the moon. "But I didn't work on 13, as much as other flights, because I was getting ready for Skylab, which was my last mission," he says.

Bales remained with the space program until 1996, eventually working his way up to the position of deputy director of operations with 9,000 workers under his supervision. "Everything paled next to the moon missions for a while," he says, "but I was still a young fellow, still very narrow. I knew this one system, this one little thing—the lunar module's guidance system. So as I got older and matured a bit, I got jobs with bigger and bigger scopes."

Bales left NASA to help direct Amspec Chemical, a New Jersey–based company that makes fire-retardant additives in plastic household products, such as televisions and appliances. "I figured if I was ever going to do anything else, it was time to make the move," he says. "I wasn't leaving NASA in a lurch. The people behind me were fully qualified to take over."

He married in 1984, and for the longest time, his wife Sandra and their two children didn't know what a critical role he had played on Apollo 11. "Number one, I don't like a lot of attention," he explains. "And number two, not too many people know about me, so it's not brought up a lot."

While Bales doesn't talk a lot about his NASA days, some of his most treasured keepsakes are from the lunar landing missions: two flags that flew to the moon; his personal notes and a tape recording of the Apollo 11 descent; a team photograph of mission control.

His fifteen minutes of fame occurred in California, shortly after Apollo 11's triumphant return. When President Nixon presented Armstrong, Aldrin, and Collins the Medal of Freedom, Bales received one, too.

"He certainly deserved it," Aldrin was quoted in *Apollo Expeditions to the Moon*, edited by Edgar M. Cortright for NASA, "because without Steve, we might not have landed."

"I accepted that award for all the thousands of people who worked on the Apollo program," Bales says. "And, yes, all the young guys who came in and helped make our nation's dream a reality."

2 BRUCE McCANDLESS

Apollo 11 Capcom

The photograph has been used in countless magazine advertisements for an obvious reason: It is a guaranteed page-stopper. An astronaut is shown floating freely in space, nothing connecting him to his spacecraft. His backdrops are the deep black void of space, and a brilliant blue earth and its atmosphere, which looks like nothing more than a thick layer of fog.

It is Bruce McCandless. He flew on two shuttle missions, made the first untethered spacewalk, helped deploy the Hubble telescope, and was inducted into the Astronaut Hall of Fame in 2005. But he is best known to the general public for his voice. McCandless was the capsule communicator in mission control who talked Apollo 11 astronauts Neil Armstrong and Buzz Aldrin through history's first moonwalk. "*OK, Neil, we can see you coming down the ladder now.*"

"When they were younger, my [two] children would recognize that it was Daddy's voice, when something was shown on television about Apollo 11 and the moonwalk, and [they would] really get a kick out of it," McCandless says, chuckling. "But that was Neil's and Buzz's show. They were on the moon; they knew what they were doing. Looking back, it's a privilege to have been a part of it." Typical of most ground-support personnel, he downplays his role.

But that night, in the few hours between the landing and the moonwalk, McCandless couldn't help but wonder: Would he ever have the chance to be where Armstrong and Aldrin were?

McCandless, born June 8, 1937, in Boston, was a member of the third crop of astronauts recruited by NASA—the 1966 class that dubbed itself

BRUCE McCANDLESS

Bruce McCandless, who served as capcom during the Apollo 11 moonwalk, never flew an Apollo mission. But during the shuttle program, McCandless helped deploy the Hubble telescope and made the first untethered spacewalk. He was inducted into the Astronaut Hall of Fame in 2005. [Courtesy NASA]

Bruce McCandless, who never got a seat on an Apollo flight, performed the first untethered spacewalk. Here, he floats about 300 feet from the shuttle. McCandless and the shuttle were traveling more than 17,000 miles per hour. [Courtesy NASA]

the "19 Originals," a good-natured jab at the 1959 Mercury pioneers known as the "Original 7." He came to the Manned Spacecraft Center in Houston with impressive credentials: second in his class of 899 at the U.S. Naval Academy in 1956; electrical engineering degree from Stanford University; fighter pilot aboard the USS *Enterprise* as part of the Cuban blockade in 1964.

The other eighteen had dazzling resumes, too, which didn't bother McCandless. "Nobody believed that Apollo 11. . . . Let me rephrase that. Nobody believed early in Apollo that the fifth flight would land on the moon and come back successfully. Everybody figured the program would take much longer, and I think they selected us with that in mind."

When a fire in January 1967 killed Gus Grissom, Ed White, and Roger Chaffee during a launch simulation for Apollo 1, and stopped the program for eighteen months so the command module could be overhauled for safety purposes, McCandless was certain Apollo would be a program of tiny steps that eventually would lead to the moon.

But like most others, he failed to understand that the Apollo program seemed only to gain momentum from any setback. It was like a boxer who, when punched once, would punch back twice.

"When things came together, and we were finally ready to fly [with Apollo 7], things went at break-neck speed," McCandless says. "The second manned mission [Apollo 8] went around the moon, which was the first manned flight of the Saturn V launch vehicle. I don't know in today's environment if we would be that bold."

He watched curiously as others in his class earned backup crew assignments: Fred Haise on Apollo 8 and 11; Ed Mitchell on 10; Jim Irwin and Al Worden on 12. Once an astronaut served on a backup crew, a prime crew assignment was almost a certainty. But even though he hadn't served in a backup role, the numbers remained in his favor. The program was scheduled to go through at least Apollo 20—nine more flights, twenty-seven more seats.

"Stay focused, do your job, and your opportunity will come," McCandless kept reminding himself. "Why would they have picked all these astronauts and invested so much time and money training them, if they didn't intend on sending them up?"

So McCandless did what he was told by Deke Slayton, director of flight crew operations and in charge of selecting the astronauts for each mission, and Alan Shepard, chief of the astronaut office. He gladly accepted the assignment as Apollo 11 capcom. He already had been named capcom for Apollo 10, the dress rehearsal two months prior to the actual landing. McCandless looked at his involvement in these historic flights as votes of confidence from the two men who mattered most when it came to getting a seat to the moon.

"The crew had some say-so and preferences on who the capcoms would be," McCandless says, "and those were frequently honored. I think Al Shepard and [Apollo 10 commander] Tom Stafford combined to pick me for 10. And then with Apollo 11, it was a matter of maintaining continuity and carryover.

"For me personally, I was in the mix," he says. "At that point, that's about all you could ask for."

By late 1968, the Apollo program was full speed ahead. President Kennedy had promised a man on the moon by the end of the decade, and NASA was determined to deliver.

"We started launching flights every two months," McCandless says. "And the training load and the development load were absolutely horrendous. I'm not sure we could've kept up that pace had Apollo 11 not been the success it was."

In October, Apollo 7—with Wally Schirra, Don Eisele, and Walter Cunningham aboard—evaluated every facet of the command module during the ten days, twenty hours, nine minutes it spent orbiting the earth. It performed so well, Apollo 8 took the command module for ten orbits around the moon two months later. Again, it was nearly flawless.

A huge question still had to be answered: How would the lunar module, a highly weight-conscious craft designed solely to fly two astronauts to the lunar surface and back to the command module, perform in the weightlessness of space? Any glitch could push the program back months. Apollo 9—guided by Jim McDivitt, Dave Scott, and Rusty Schweickart—silenced concerns in March 1969. The LM proved easily maneuverable, docked smoothly with the command module, and was solid during rendezvous exercises.

In May, Apollo 10 lifted off to perform even more intricate tests of the lunar module, as it orbited the moon, while the Apollo 11 crew tirelessly trained for the landing. Suddenly, Bruce McCandless was as busy as any astronaut on either flight. "I can remember days when I'd serve four-hour shifts as capcom during the flight of Apollo 10, take a lunch break, then move over to the other control room for a four-hour shift as capcom for Apollo 11 in training. You really had to keep mental processes straight so you understood what was real and what was a simulation exercise. I know that sounds strange, but that was life around the control center at the time."

Again, the lunar module—*Snoopy*—showed the right stuff on Apollo 10, dipping Stafford and Gene Cernan to within nine miles of the jagged lunar landscape at speeds approaching 4,000 miles per hour. Valuable photographs were taken of landing sites, especially Apollo 11's target in the Sea of Tranquility. The LM's radar, attitude control, propulsion, and communication systems all checked out. The only hint of a problem occurred when *Snoopy* began on its path to rejoin John Young in the command module. Its automatic abort guidance system sent the spacecraft tumbling for about eight seconds. After countless evaluations by engineers, it was deemed a mere hiccup; something they could easily tweak.

Next step: The Sea of Tranquility, Moon.

Capcom duties were not easy. The position almost always fell to a fellow astronaut because the guys in space felt more comfortable communicating with one of their peers. And unless it was an extreme circumstance, the capcom was the *only* person in mission control allowed to speak directly to the astronauts.

"First off, one of the prerequisites for being a capcom was that you had to train with the specific crew you would be working with so that you knew what they were thinking, what sort of idioms they might be using, when to butt in, and when to remain quiet," McCandless says. "You also tried to avoid over burdening the crew. [You had to] make sure whatever information was being sent up—particularly numerical data—was understood correctly so you didn't wind up with a screw-up in the computer."

Strict protocol among members of the flight team, who were plugged into the communication system inside mission control, was a must. "People coming onto the [communication] loop were supposed to state who they were and say what they needed to say in a clear, concise manner. For instance, 'Capcom, this is surgeon. Tell them not to work so hard on the lunar surface. Their heart rate is going up.' Something like that. The flight director was there, and if he disagreed with what the capcom was being told to send up to the crew, he stepped in and said, 'No, hold on a minute.'

"One of the real assets of the hard training was getting to know everybody, learning their voices, and knowing who was talking to you at all times. Because at any given moment, you might have three or four people in your ear all at once, trying to get their information up to the crew."

Training for Apollo 11 was a whole new assignment. No one had ever been a capcom for a moonwalk. And the EVA simulations, performed in Florida's sandy soil near the Cape, with a few volcanic rocks scattered about for the astronauts to analyze and document, were never performed start-to-finish. Instead, they practiced much like a movie is filmed, in segments. "So it was hard to get a real feel for the thing," McCandless says. "It was a matter of doing several part tasks, then mentally piecing them together."

The deeper he got into training, the more McCandless understood the logistical nightmare ahead of him. Not only did he have Armstrong and Aldrin to communicate with on the surface, but he was also responsible for relaying the normal flight-related information to Collins who was orbiting the moon in *Columbia*. He felt a bit like the director of a Broadway play, with an outline but no script and a curtain that was about to go up.

Plus, there was the historical significance of the event to consider. More than 700 million people worldwide would be watching Armstrong and Aldrin make tracks in dirt billions of years old. Surely, Armstrong had thought of this and would have something memorable to say, as he reached the bottom of the ladder to put the first human print on another world. And the last thing McCandless wanted was for his voice to trample over some historic prose. This was a one-take production. So he asked Armstrong numerous times what he had planned. Armstrong refused to say. He would tell McCandless: "I'll probably just say something like 'Boy, it sure is dusty up here. OK, here we go. . . . There's a rock.'"

Says McCandless, "It became a point of contention between Neil and me. I felt like I needed to know. He didn't see it that way. The one thing I was sure of was that I had decided I was going to say as little as possible during the first portion of the EVA. My thoughts, my voice [were] readily available. But they were the guys on the moon; they were the guys with the time constraints, in terms of gathering rocks and taking pictures and setting up equipment. They were given a license to innovate and modify things, and we'd just see how it went."

Late on the afternoon of July 20, 1969, as Armstrong demonstrated why he was selected for Apollo 11 by guiding *Eagle* over and through boulder fields with expert precision, McCandless, off duty, was sitting in the crowded control center on a carpeted ledge that normally served as a bookcase. Charlie Duke was the capcom for the landing.

"There were so many extra people around, you could barely find a place to plug in a headset," McCandless recalls. "When they landed, there was all this cheering, and everyone was really happy when it was verified that everything was in great shape."

Everyone, that is, except the geologists. Because Armstrong had to fly *Eagle* longer than anticipated to find a suitable touchdown area, no one could be sure exactly where it was on the moon.

"Prior to the landing, the [geologists] had taken [the] position: 'We don't care where they land specifically; we just want a safe landing, and we want them to get their rocks and soil samples.' And everybody agreed that was the reasonable approach," McCandless says. "Within thirty seconds after touchdown, they were screaming, 'Where are they? Tell us *exactly* so we can maximize their time on the surface!' The truth was, we didn't know. We knew somewhere within a several-hundred-yard range, but there was no way to know exactly."

In fact, this would lead to more work for McCandless. Geologists wanted Collins to look through a special telescope on *Columbia,* in an attempt to spot *Eagle*, using suggested coordinates read to him by McCandless. "I can't even remember how many sets of coordinates we sent up to Mike, hoping he might spot them as he passed over during his orbit. But he never found them. It wasn't until they returned from the moon, using photographs taken by Armstrong and Aldrin, [that we were] able to identify features and rocks and things, and pinpoint exactly where they had landed."

Soon after the landing, McCandless headed for his home, which was about three miles from the Manned Spacecraft Center. The astronauts were to eat, have a sleep period, and then do the EVA.

"No sooner had I gotten home, word came that [the astronauts] couldn't sleep, and they were going to do the EVA early," he says. "I don't think I could've slept, either. So I jumped into the car and headed back to the control center."

It remained packed with NASA officials and VIPs, but that didn't faze McCandless. Nothing did. Not even the site of Neil Armstrong, a guy he had known for three years, making his way down the ladder toward history.

As he had decided during simulations, McCandless said little during Armstrong's first few minutes of exploring. In the first thirty-one minutes of the moonwalk—during which time Armstrong greeted the moon with his left boot, uttered his famous line ("That's one small step for man, one giant leap for mankind"), and was joined on the surface by Aldrin—McCandless spoke only twelve times. Six of those were a simple "Roger" to let the astronauts know mission control was hearing them. Twice, he answered questions from Armstrong about the positioning of the TV camera. Twice, he urged Armstrong to hurry and get the contingency sample—a quick scoop of soil and rocks, in case the EVA had to be ended abruptly. Once, he told Aldrin his voice transmission was cutting out, and he asked Armstrong if he envisioned any trouble transferring the rock boxes up to the LM.

During the remainder of the first thirty-one minutes, McCandless allowed the magical scene to unfold with little disturbance.

His voice became more prominent once Collins arrived from the moon's dark side. Collins needed computer updates, but more importantly, he wanted to know how his buddies were doing down there on the moon. McCandless reported to Collins: "The EVA is progressing beautifully. I believe they're setting up the flag now. I guess you're about the only person around that doesn't have TV coverage of the scene. Oh, it's beautiful, Mike. It really is."

Later, McCandless said to Collins: "They've got the flag up now, and you can see the stars and stripes on the lunar surface."

Throughout the one hundred and fifty-one-minute EVA, McCandless was cool and smooth. The longer it went, the more he seemed to

Mission Control in Houston was jam packed during the Apollo 11 moonwalk, the first ever by humans on another world. [Courtesy NASA]

enjoy it. An hour into the EVA, he had this humorous exchange with Aldrin.

McCandless: Buzz, this is Houston. You're cutting out on the end of your transmissions. Can you speak a little more closely into your microphone? Over.

Aldrin: Roger, I'll try that.

McCandless: Beautiful.

Aldrin: Well, I had that one inside my mouth that time.

McCandless: It sounded a little wet.

McCandless never flew an Apollo mission. Budget cuts ended the program after Apollo 17. "Does it bother me?" he says, repeating the question. "It doesn't *bother* me. Obviously, I would like to have gone to the moon. But out of the whole space program, we've got twelve guys who walked on the moon and two who came close [Lovell and Haise on Apollo 13]. Actually, I think about people like Joe Engle. He was set to go on Apollo 17 but got replaced by Jack Schmitt, the geologist. If anybody has regrets or remorse, it's probably Joe. Hey, we were all ready and wanting to go."

Certainly, McCandless' achievements in the shuttle program helped ease his pain. The Hubble telescope he helped deploy in 1990 has given astronomers their deepest look into space. Photographs of his untethered spacewalk are some of the most famous in NASA's storied history, perhaps because he looks so . . . free and to himself.

"You're not going to believe it," McCandless says, "but do you know the best description of that spacewalk from my perspective? Noisy. Bob Stewart was out on a [tethered] spacewalk at the same time. We had three guys in the orbiter, plus the people on the ground. And they all seemed to want to talk at the same time." (Ironic since he made it a point *not* to talk too much during other astronauts' shining moments.)

"Sometimes people mention it and say what a profound experience it must've been," he continues. "Well, it wasn't, although as I look back on it, I very highly value it, and I did thoroughly enjoy it. It's sort of the situation where you have a trained professional who works very hard to train for something, and then you finally get the opportunity to do it. And I had trained for probably sixty or seventy malfunctions, or combinations of problems, that might [have made] it challenging to get back to the orbiter. But *nothing failed*. The unit worked beautifully. After 41-B, it was refueled and loaded back in the orbiter payload bay and used on 41-C, with no changes, no repairs whatsoever."

One of McCandless' crewmates aboard *Challenger* on STS 41-B was Ron McNair. He would perish in that same vehicle two years later, along with six other astronauts, in an explosion shortly after launch. McCandless and McNair had become close friends while working together on a cinema experiment for a planetarium in Jackson, Mississippi. During the mission, McCandless took photographs of McNair wearing a director's hat and carrying a movie-set slate. "He was a regular Cecille B. Demille up there," McCandless says, managing a laugh.

And for the next several minutes, he talks about his deceased friend; how much he admired him; what a great future he had predicted for McNair.

He speaks quietly, almost reverently, and he doesn't count his words.

3 RICHARD UNDERWOOD

NASA Chief of Photography

Richard Underwood had just finished processing the precious film brought back by earth's first moonwalkers, Apollo 11 astronauts Neil Armstrong and Buzz Aldrin, when he realized something was missing.

He hurriedly looked again through the hundreds of images taken on the night of July 20, 1969. Then he looked again . . . and again. There was plenty of Aldrin—saluting the U.S. flag, scooping up soil samples, standing on the ridge of small craters. But nowhere could Underwood find a picture of Armstrong.

"I told one of the lab technicians," Underwood recalls. "He said, 'Aw, they're in those spacesuits. They both look alike.' Then we got the film dry and looked at it. Sure enough, none of Neil. Buzz didn't take any."

It created a nightmare for NASA's public relations office. "In a meeting, there were people from [public affairs] who said, 'Look, this one of Aldrin standing by the flag? Let's just say it's Neil Armstrong. You can't see his face.'" Underwood says. "I quickly told them, 'Yeah, but there's some nine-year-old groupie out there who understands these space suits as well as the people who made them. And if you say that, three days from now there will be a letter in the *New York Times* saying, 'You're full of crap, that's Buzz Aldrin.' So they just decided not to say anything.

"And the amazing thing is, the news media never asked about it. Year after year, no one ever said a thing about it."

Underwood, who now gives speeches about the Apollo moon missions for Crystal Cruise Lines, says his audiences are stunned when he tells them there are no pictures of Neil Armstrong walking on the moon.

RICHARD UNDERWOOD

Richard Underwood was known as the astronauts' photography "coach." [Courtesy NASA]

"It's the first time they've ever heard it," Underwood says. "The news media really dropped the ball on this one."

Actually, Aldrin did take a couple of photos of Armstrong on the lunar surface, but they are out of focus, and Armstrong has his back to the camera, as he is working at the lunar module's equipment bay. He looks about the size of an ant.

And there is one other image of Armstrong on the moon. In what many consider the most famous picture taken during any moonwalk—a full-length, straight-on shot of Aldrin standing on uneven ground, the legs of his white spacesuit smudged by gray moon dirt—Armstrong is slightly visible in the reflection of Aldrin's gold-plated visor.

In countless interviews, Aldrin has said the snub of Armstrong was not planned. Aldrin blames it on a tight timeline and says Armstrong had the camera most of the time.

Underwood, NASA's chief of photography during Mercury, Gemini, and Apollo, doesn't buy it. "Buzz took plenty of pictures while on the moon. He just didn't take any of Neil," he says. "So in the thousands of photographs we have from the six lunar landings, we have none of the first person to walk on the moon. It's sad. And, in my opinion, Buzz Aldrin didn't take any of Neil because he was pissed off about [Neil] being the first to walk on the moon."

For years leading up to the moon voyages, Underwood had fought to make photography a priority among mission planners.

"Some of the engineers were against taking pictures at all," Underwood says. "They said it wasn't worth the weight to take a camera aboard. And they argued that the astronauts were too busy flying the spacecraft to take pictures—and they were right about that, to an extent. But I used to scream at the engineers in meetings that the key to the immortality of the whole project was based on photography. I told them, 'You're going to spend $50 billion on everything else, yet you don't want to spend $20,000 on cameras—and pictures are the total program? Without those pictures, we'll have no idea what happened up there. You can load thousands of books with all this computer data about our trips to the moon, shove them in a library, and nobody will ever read one.'

"And the last thing I used to tell the astronauts, before they flew, was the key to your immortality is your photos. Take good pictures."

For the most part, the astronauts listened. But in the early 1960s, as NASA launched its step-by-step plan to land a man on the moon by the end of the decade, no one was sure it was even possible to take pictures in space. Tests performed by Underwood and his staff revealed numerous problems.

"Film was the main one," Underwood says. "The film you use on earth turns to powder in space after a while. With the low humidity, the low air pressure, and the low temperatures, the film breaks down.

"The leatherette covering that was on most cameras back in the early days of spaceflight looked impressive and had a nice feel to it, but in space, it would outgas and asphyxiate the astronauts. If you peeled the leatherette off, you had a camera made of stainless steel. Sunlight coming through the window of a spacecraft and bouncing off would permanently blind the astronauts—and we didn't think blind astronauts [was] a good idea. It wouldn't sell the program real well.

"And the springs that operated the focal-point lens wouldn't operate properly in zero gravity. So here we are, about to go into space and to the moon, and we're going down this blind alley as far as taking pictures was concerned. But space photography evolved very quickly. We had to come up with some things that would work."

In 1963, Underwood went to the world's most recognized camera experts for help. Victor Hasselblad's great-grandfather started a small camera company in 1841. By 1890, FW Hasselblad and Co. was the Swedish importer for Eastman Kodak and on its way to worldwide fame.

In just three months Hasselblad's engineers had developed a leaf-shutter camera that would work in space, and engineers from Eastman Kodak came up with space-proof film.

"The film had to be coded differently for space," Underwood says. "With any aerial film, there is a big problem with the blue [sky]. With normal film, you record your problem first, which is the blue haze, and if there is any light left over, your subject second, which is the earth. We wanted film that would capture the subject first and the problem second.

"The film they came up with was very thin. The emulsions were placed on [it] differently. It was about as thin as Saran Wrap, which meant you could load a lot of it into a single magazine. Astronauts didn't have a lot of time to screw around with changing magazines. Plus, it was virtually impossible with those bulky gloves on."

Underwood, born in Rhode Island on July 15, 1926, had always enjoyed solving problems concerning photography. While earning degrees in civil engineering and geology at the University of Connecticut, Underwood wrote a paper about how infrared film could help solve a certain geological task.

"I'd heard during World War II that we had infrared film that could see through the German and Japanese camouflage, and I wrote a letter to the research department of Eastman Kodak [in 1950] and asked them about it," he says. "I rented a piper cub, a friend of mine flew it, and I graphed the area that had been assigned to me and solved the problem."

While at University of Connecticut, Underwood was staff photographer for the student newspaper, shooting football games and award ceremonies. "Interesting, but not very challenging," he recalls.

Once out of college, Underwood went to work for the U.S. Corps of Engineers as a worldwide aerial photographer. Most of the time, two cameras were onboard the B-17: Underwood's, which was used for mapping, and another for gathering intelligence. As spies in the skies, Underwood and his crewmates often drew fire from Russian MIGs.

The enemy never managed to ground him, but romance did. While on assignment in Honduras in 1954, Underwood met and married a young girl named Rosa. "I immediately lost all my security clearances," he says. "They pulled me off the high planes and sent me to Florida on an unclassified project."

Underwood didn't know it at the time, but he had just joined the space program. His new boss was Wernher von Braun, the architect of the rockets that, eventually, would win the space race for America.

"They were about to begin test-firing rockets from Cape Canaveral, and I was to help hang cameras on the missiles so we would know where they wound up landing, and [so] we could perfect the guidance systems," he says. "Von Braun was a natural-born leader, the type of individual who makes you want to do exactly what he wants done. I haven't run into too many people like that."

In twenty-seven manned space missions, there was never a major photographic glitch. Underwood didn't always get the exact picture he wanted, and there was the Aldrin snafu on Apollo 11, but the cameras and film almost always worked. And the pictures, particularly from the moon, are high quality, especially considering they were taken by astronauts whose demanding training schedules didn't allow much time for learning the nuances of photography.

"We worked hard to make sure the cameras were astronaut proof," Underwood says. "On the moon we eliminated all the problems for them. The camera [a Hasselblad 500EL] was set. The light was constant. Just look out there, find a good scene, and take a picture. It was basically point-and-shoot.

"They could change the [camera] settings, if it was an emergency—if they wanted to shoot something in the shadow of the spacecraft, or something like that. But they had a lot to worry about, and we didn't feel like changing f-stops should be one of them.

"The cameras were battery powered and had to have automatic drive. The astronauts would've had a hard time cranking the film manually with those big gloves on. The button they had to push to take a picture was about a square inch [in size] and [was] located just below the lens on the right side. Each roll of film was 4,800 feet long and would hold between 300 and 400 pictures."

Underwood and crew knew every switch and screw inside the cameras. They proved this on Apollo 12 when command module pilot Dick Gordon, orbiting the moon alone as crewmates Pete Conrad and Alan Bean walked on the surface, reported his camera was jammed.

Says Underwood, "We'd run every possibility through a computer and asked: Does the astronaut have time to fix it? The equipment to fix it? The smarts to fix it? Some problems would come back: 'He can't fix it.' But when Dick told us what was happening, we determined that it was the last-picture switch that was malfunctioning. When you took the last picture on a roll, it would shut the system down automatically so you weren't shooting pictures with no film. So we figured the switch had dislodged.

"We told him to take a screwdriver and remove a little plate covering the switch. There were two screws holding it in place. We said to take the first screw out and put it in his mouth—we didn't want it to float around, then get stuck in something—then take the second one out and do the same thing. Then we told him to pry up the corner of that little plate, and a little screw and a little wire would come floating out.

"Dick told us we were nuts. But when he pried it up, I heard him say, 'Well, I'll be.' We had him take the switch out, and get rid of it, and told him we'd keep up with how many pictures he'd taken on each roll from the ground. On our monitors in mission control, we got a little blip on a screen each time a picture was taken."

Pictures on each mission were carefully thought out and written into the flight plan. On Apollo 8, the first trip into lunar orbit, the plum photos weren't hard to figure out: the first images of earth, totally suspended in space and decreasing in size as the command module raced toward the moon, and earthrise as the astronauts rounded the backside of the moon.

"All three astronauts realized it was an historic mission," Underwood says, "but there were some [time] battles going on within NASA, especially during lunar orbit. I argued hard for the shot of earthrise, and we had impressed upon the astronauts that we definitely wanted it. I think the surprise of the whole thing was the beauty of it. Maybe they weren't ready for that. How could they be?

"Bill Anders wound up taking the picture. I had a pretty good idea of what it was going to look like, but when I actually saw the picture, after they returned, it was even better than I had anticipated."

The photos of a diminishing earth didn't turn out as well. In fact, not until Apollo 17, the last of the nine moon missions, did Underwood get the series of pictures he wanted.

"Jack Schmitt was a geologist and geo-physicist, so he understood the essential values of pictures of the planet earth as you moved away from it," he says. "What we wanted was a record of what parts of the earth are visible, how far out certain things disappear, the movement of the earth itself. We put it together as a time-lapse thing.

"Astronauts on the other flights would take pictures of the earth periodically, but not a series of the earth diminishing."

Richard Underwood says this is the best photograph ever taken from space. It shows a full earth and the continent of Africa, along with the polar ice caps. Photo taken by Jack Schmitt, Apollo 17. [Courtesy NASA]

Apollo 17 also delivered the panorama of the moon's surface Underwood had craved. "I felt they were valuable because a panorama would be just like being on the moon. We'd have the astronaut take a picture, turn thirty degrees and take another, and keep going until he had gone all the way around. Then we could overlap them, show them in a circular room, and the scientists could stand there with the astronauts twenty years later and say, 'Now what was that over there?' Neil took a partial panorama on Apollo 11, but we finally got a good one on Apollo 17, because they had more time with three EVAs."

Underwood still shakes his head over a decision made by geologists before the November 1969 flight of Apollo 12. Conrad and Bean were asked to take pictures of Surveyor 3, an unmanned U.S. vehicle that had landed on the moon in April 1967—but only in black and white.

"I argued, but that was a battle I lost," Underwood says. "The scientists said there would be more information available in black and white pictures. I suggested one astronaut take black and white, the other take color. They kept saying, 'No, no, no, we want them all black and white.' It was an idiotic decision by some guys with three PhD's apiece."

Some of the most valuable images were taken on Apollo 13, the flight that never reached the moon's surface. Command module pilot Jack Swigert was able to photograph the jettisoned service module, where an explosion had occurred, while it was on the way to the moon. It nearly cost Swigert, Fred Haise, and Jim Lovell their lives and ruined the mission. The pictures showed one side of the spacecraft, nothing but a spaghetti-tangle of wires and pipes. Swigert's work helped engineers evaluate what had happened and why.

And those photos were much more than point-and-shoot. "We weren't sure they'd be able to get a picture of it," Underwood says. "[Astronauts] John Young, Ken Mattingly, and I went into the simulator, and our first inclination was no. Then we got into a lot of things. There had to be changes in the rotation of the spacecraft. [The command module constantly rotated so that no one side faced the sun too long.] Then we had to figure the f-stop, because we knew the sun was going to be facing us.

"Jack made the shot with no viewfinder and a 1000 mm lens that was literally, a yard long. We had to countdown the precise second the service module would appear in the window. Jack couldn't look out the window and take the pictures because the lens was so big. So in the simulator, we figured out where the lens had to touch the window, where the back end of the camera had to touch the ceiling. And we knew he'd have time to take three pictures. It just happened to work out."

Underwood says the best photographers among the moon voyagers were Young, Armstrong, Conrad, Schmitt, and Tom Stafford. "They took every task seriously," he says. "But the quality of the photographs varied from human to human, just like down here. Some were quite anxious and realized the value of the photography, not only as a record, but [also] to open new doors in science and engineering. And then others were quite aloof about it."

Ironically, Underwood's favorite photo from the lunar surface was taken by Aldrin; it was a shot of his boot print in the lunar dust. But the most meaningful to him personally was a photo taken by Gene Cernan on Apollo 17 of the U.S. flag planted on the lunar surface.

"My mother gave me a St. Christopher in the spring of '44, and I wore it all through the war as I loaded and unloaded ammunition ships," Underwood remembers. "That medal went on all the previous flights, all the way back to Mercury. Gus Grissom carried it in his boot. They'd take it into space, then come back and certify that it had flown."

Richard Underwood's favorite photo from the moon: a boot print taken by Buzz Aldrin on Apollo 11. [Courtesy NASA]

On Apollo 15, Underwood's St. Christopher wasn't the only memorabilia to make the trip. Astronauts Dave Scott, Al Worden, and Jim Irwin took canceled stamps that would be sold by a German collector upon return, with the crew due to get a sizeable cut. When NASA heard about the stamps, it publicly reprimanded Scott, Worden, and Irwin, and a new mission rule was written: Only items from the crews' immediate families would be allowed to fly.

Underwood knew his St. Christopher wouldn't be aboard for the final trip out to the moon.

"In the last briefing at the Cape before Apollo 17, I asked Gene and Jack if they would do me a favor," he says. "They said, 'Sure, what do you want?' I said, 'Will you make sure the last picture taken from the moon will be of our country's flag?' They said sure. And they did. You can hear on the audiotapes, just before takeoff, Jack asking Gene, 'Did you get Dick's picture?' He said, 'I sure did.' And everybody in mission control started looking over at me, but I didn't really care."

Today, Underwood keeps the St. Christopher in a safety deposit box.

Every person involved in the moon missions had their own pressures to deal with, and Underwood was no different. He was in charge of developing procedures for handling, processing, and preserving the film used by the astronauts. There was no room for error.

"Once the astronauts got back into the lunar module after the EVAs, the film was stored in bags, similar to what they put the moon rocks in," he explains. "When they docked with the command module, they were transferred over, and the command module pilot checked off each bag as they were handed to him.

"After splashdown, and once they reached the aircraft carrier, the film would be removed by our guys in the photo department and inventoried in a special area. Then [it was] placed in a special container, sealed, and transported back to Houston. [It] landed at Ellington Air Force Base, about five miles from the space center, and it was brought into the lunar receiving laboratory. We had pictures only a few hours after they got back."

Terry Slezak, one of Underwood's technicians, became the first non-astronaut to be exposed to moon dust. "When the first batch of film was brought to us, [Terry] pulled out the first magazine and read aloud the [catalog] number on it. Then he transferred it to his other hand, and we could all see that the hand he'd been holding it with was black with moon dust. We started kidding him, 'In ten seconds you will self-destruct. Thanks for your good work, Terry.' One thing we found out in a hurry is that moon dust sticks to everything."

So Slezak quickly was sent into the quarantine trailer. Along with the Apollo 11 astronauts, he spent the next three weeks making sure he was not contaminated with some human-destroying bacteria.

"I kept saying we should leave the film in the quarantine trailer with the astronauts, but NASA wanted pictures right away," Underwood says. "So we spent two years coming up with a plan to decontaminate it. We figured out that ethylene oxide, when used as a gas at a certain temperature, would, in theory, kill any pathogenic device that might be on the film. So that's what we did—stuck the film in containers with ethylene oxide for a few hours."

Master copies were made of all the original film. Copies were then made from the masters. "The original film from every mission is stored in stainless steel cans, pressurized, and frozen," Underwood says. "We eliminated [the possibility of] any oxidation. The film should be there forever, so long as some idiot doesn't make the decision to destroy them all."

Few people know just how close the world came to never seeing the pictures from Apollo 11.

"The spacecraft was about to splash down, and we were running through one final test on the film processor, which had been checked hundreds of times before," Underwood says. "When we opened the container, the film had melted. It was just one big glob of gook. We discovered at that point that somehow the ethylene oxide was dripping onto the film. So we started scrambling, knowing that the first moon film was about to be delivered to us, and we built a stainless steel cover to prevent it from dripping onto the film.

"It was just pure dumb luck that we decided to do one more test on that processor. Had Armstrong's film been put in there without that last test, it would've eaten it up. It would've been the greatest photographic catastrophe in the history of the planet. Thank God it didn't happen."

4 CLANCY HATLEBERG

Frogman, Apollo 11 Recovery

Clancy Hatleberg could easily have been in Vietnam doing one of his many stints as a U.S. Navy underwater demolition specialist; clearing rivers and canals of bombs; performing reconnaissance. "You know, James Bond type stuff," Hatleberg says with a laugh.

Instead, that summer of 1969, when Apollo 11 carried the first humans to walk on the moon and was scheduled to splash down in the Pacific Ocean, by luck of the rotation schedule, Hatleberg was back on the West Coast. Hatleberg's outfit—coincidentally known as Underwater Demolition Team 11—drew recovery duty, working off the USS *Hornet*, an aircraft carrier that launched numerous bombing attacks during World War II.

"When I first got the orders, I was tremendously excited because I had always believed in the space program," Hatleberg says. "Being given a chance to participate in the first mission where men would actually walk on another planet. . . . It was like a dream come true."

But he quickly realized this recovery presented serious new challenges for NASA, even for the world. What if, scientists wondered, the astronauts picked up deadly bacteria during their stay on the moon that would thrive once subjected to earth's atmosphere? As silly as it may seem, moon travelers Neil Armstrong, Buzz Aldrin, and Michael Collins had to be treated as if they were bacterial time bombs capable of destroying human life by simply exhaling.

"People were very, very concerned about an H. G. Wells, *War of the Worlds*, sort of in reverse," Hatleberg says. "Instead of defeating the Martian invaders, we could possibly be hit by a load of diseases.

CLANCY HATLEBERG

Clancy Hatleberg, a former navy scuba diver in Vietnam, with his daughter Jodie, a firefighter in San Diego. [Courtesy Clancy Hatleberg]

"Scientists were worried that if there was something on the moon that could survive all the radiation and survive the vacuum of space for billions of years, then it might be extremely potent if exposed to, say, water. Then what would we do? We've seen what things like fevers and HIV can do, how quickly they can spread. How could we be sure there wasn't something up there hundreds of times more potent?"

NASA engineers and scientists were confident they had developed a new recovery plan that would prevent any chance of the astronauts spreading germs into the atmosphere. "The recovery team was to secure a flotation collar around the command module. Then the recovery ship would come alongside and use a crane to lift it out of the water and onto the hangar deck," Hatleberg explains. "Then the astronauts would walk from the command module through a cocoon tunnel and into a quarantine facility."

One problem: When the Apollo 1 crew died in the fire, investigators ruled the spacecraft contained too many flammables. Changes were made. The new and improved command module was much heavier—*too* heavy for the crane that was to lift it from the ocean.

"Under some sea [conditions], if the command module was going down with the water, and the crane was pulling up, you could rip the hatch right off. And then you had all sorts of problems," Hatleberg says.

Because the plan to use a crane appeared to be flawed, the astronauts would have to be removed from the spacecraft, hoisted up into a helicopter, and delivered to the ship's deck—just as in previous missions. But with the seal of the spacecraft broken, as the crew egressed, and the astronauts coming in contact with swimmers, helicopter crew members, and some people onboard the *Hornet*, how could the dangers of contamination be controlled? While Hatleberg's crewmates were successfully snagging Apollo 10 from the Pacific, after its successful dress-rehearsal flight around the moon in May 1969, he and a team of NASA scientists and engineers were in Houston, hashing out an acceptable recovery plan for Apollo 11. They had about eight weeks to develop it, test it, and write a detailed procedure.

Hatleberg shrugged and looked at the assignment this way: It beat the hell out of what some of his buddies were being asked to do in Vietnam.

Hatleberg was not the first member of his family to serve his country in the Pacific. His dad, the late Earl Adrian Hatleberg, was a Marine regimental surgeon—on loan from the Navy—and one of the first to hit the beach at Iwo Jima during World War II.

"He never talked much about it," Hatleberg says, "but he did say that he established a hospital in the sand, and that when his buddies got shot, all he could do was inject them with morphine. That bothered him all his life. He would say, 'That's all I could do.' I think his words were some sort of strange confession."

After the war, the Hatleberg family settled in Chippewa Falls, Wisconsin, where Earl had been stationed for a while. Clancy grew up skiing year round, on snow or water. "I actually harbored dreams of becoming an Olympic downhill skier," he says. "But then I met these things called moguls, and reality kicked in."

His life changed on a summer day in 1958 at Lake Wisota, a water-skiing hot spot near Hatleberg's home. It was there he first noticed Susan Claire Erickson.

"I was fifteen, she was fourteen, and it was one of those rare moments that you just never forget," Hatleberg says. "It was early evening, so the lake was dark and smooth. She was wearing a black bathing suit, and her brown hair was in a ducktail. I had never seen a girl like this before—so beautiful, so athletic. She was slalom skiing, and she was *good*.

"About a year later, I got my driver's license and drove about twelve miles down to Eau Claire High School to pick up a friend, and she asked if I could give her friend a ride home, too. I couldn't believe it when I saw her. It was the girl I had seen skiing that day."

Clancy and Sue became high school sweethearts, and the relationship continued even after she went away to the University of Minnesota and Hatleberg joined the Navy ROTC program at Dartmouth College in Hanover, New Hampshire. "But when I came home the summer before my junior year, she broke up with me," Hatleberg says. "My heart was shattered, so I figured I'd do what all jilted lovers do—join the French Foreign Legion.

"I actually called up the French Embassy and came across just like I was—a little naïve. The gentleman said in this thick French accent: 'Zee Le-schand eez not accepting eeny more applikaschunz.' What a bummer.

"So then I figured I'd do the next-best thing and become a Navy frogman."

In high school, Hatleberg and a few friends saved their money and bought scuba kits from Sears & Roebuck. "We'd go down to the fire station and fill up the [air] bottles and dive in lakes around home," he says. "The kit had this little manual that said, 'Don't come up any faster than your smallest bubbles.' Of course, you couldn't stay down long enough with that equipment to get the bends, anyway. But it was valuable in the sense that it allowed you to realize whether you really liked diving or not. This is not to demean any divers out there, but diving doesn't take a lot of skill. But it does take a certain amount of courage. Once you get into the water, it closes in on you real fast, and it starts getting cold, and it's not a whole lot of fun for most people."

After declaring for the Navy's diving program, Hatleberg was shipped out on his first overseas cruise. One of the stops was Toulon, France. When he walked off the ship there, he was stunned to find Sue waiting for him.

"She was in Europe traveling with a couple of friends, and she wrote my mother to find out which port I was coming into," Hatleberg says. "That was it. We were married in 1965."

While the frogman gig successfully gained the attention of his lost love, he still had to deal with Vietnam. "It was some of the worst human conditions imaginable," he says of the war. "We'd get ambushed on the river, . . . but I'm happy to say that all of us in my group made it home."

As a Navy diver, Hatleberg had grown accustomed to demanding officers and having to pay strict attention to detail. NASA took everything he had learned to another level. He was amazed by the space agency's energy and relentless pursuit of perfection.

"We trained like I'd never trained before," Hatleberg says. "We started working on Apollo 10 about a month after we got back from Vietnam. The astronauts were there just one day, but we trained in San Diego Bay with the same flight crews that we would actually go out and do the recovery with. We used this big boilerplate as a mockup of the command module.

"First time we went out, I said, 'Hey, we've got this down pat.' But the helicopter crews had done some recoveries before, and they knew better. They told me, 'Nope, we're going to keep doing this until the day we deploy.' And we still had a whole month to go!

"But all of a sudden you get caught up in the NASA attitude. You realize that these people are *really* serious about what they're doing. You were briefed before a practice procedure, and you were debriefed afterward. Any little thing you could improve on, or if you did any little thing wrong, it was all talked about and scrutinized. We did this every single day, twice a day. And at least once a week, we'd go out and practice at night.

"When you're doing that much training, it finally hits you why they're putting you through this," Hatleberg continues. "They're trying to squeeze every ounce of emotion out of you because emotion could be a distraction should something go wrong."

The recovery had always seemed the most benign part of any mission, especially one to the moon. It was only human to breathe a sigh of relief once the spacecraft could be spotted floating beneath three full parachutes, and a team of skilled Navy frogmen and helicopter crews waited to pluck the astronauts from the ocean. But during a recovery operation, the United States almost lost its second man in space. A hatch blew open, following Gus Grissom's suborbital Mercury mission in 1961, and the *Liberty Bell 7* spacecraft quickly filled with water and sank to the bottom of the Atlantic.

Grissom nearly drowned.

Hatleberg liked the new recovery plan they had developed. It was solid. As Apollo 11 sped home from the moon, he felt confident his crew was well schooled and ready for any situation that could arise. And he was vividly aware of the enormity of his mission: Deliver three new heroes and their forty-four pounds of precious moon rocks safely to the deck of the *Hornet.*

Then just twenty-four hours before splashdown, Navy meteorologists spotted a typhoon developing in the prime landing area near Pago Pago, in the Pacific. Even in ideal weather conditions, a spacecraft bobbing in the ocean never made for an easy capture. "The command module acts like a small sail in the water and can pick up enough speed to make it impossible to catch or work around," Hatleberg says.

Landing in or near a typhoon was out of the question. So the target area was switched some 200 miles to the north. The *Hornet* would have to move fast to make it there in time for the dawn landing on July 24. Hatleberg and the other members of the recovery crew would go on ahead, aboard three Navy helicopters.

They arrived in the landing area well before sunup. Inside his helicopter, Hatleberg remained calm and confident. He repeatedly went over a mental checklist of the procedure and even took a moment to remind himself how fortunate he was. He thought of his buddies in Vietnam and wondered if they'd had time to keep up with the world-changing events of the past eight days. But he was jolted back to the task at hand by a white-bright object with a long tail of fire streaking across the reddening sky. It was *Columbia*, right on time.

The spacecraft hit the water traveling thirty-one feet per second and slowly turned upside down in the choppy, blue water. The astronauts flipped a switch that inflated three flotation bags located in the nose of *Columbia*. In a matter of moments, it was upright, but far from stable.

The divers moved into position. John Wolfram, who had worked the Apollo 10 recovery, was first into the water. He attached a sea anchor around *Columbia* to prevent substantial drifting. Mike Mallory and Wes

Chesser followed. They helped Wolfram secure a flotation collar around the spacecraft, and two rafts were lowered into the water.

Now, it was Hatleberg's turn. As soon as he entered the water, he was sent down a bag of four Biological Isolation Garments (BIG)—one each for him and each of the astronauts. "I always like to say I had a big bag of BIGs," Hatleberg laughs. The dark gray garments, made of lightweight cloth, looked like baggy wetsuits. And Hatleberg's was slightly different than those donned by Armstrong, Aldrin, and Collins. Each BIG had two breathing filters around the mouth area. Hatleberg's would filter the air he breathed in; the astronauts' BIGS were designed to filter the air breathed out.

Mallory, Chesser, and Wolfram boarded one of the rafts and moved about 100 feet upwind—far enough away, if there was an extensive break in containment, but close enough in case of an emergency.

Hatleberg put on his BIG and went to work. "As soon as I was in position, standing on the flotation collar, the astronauts were radioed to open the hatch," he says. "It opened, and I dropped in the bag of BIGs.

Clancy Hatleberg waits in the Pacific Ocean with the Apollo 11 astronauts for the recovery helicopter. They are wearing Biological Isolation Garments, in case they brought back a pathogenic disease from the moon. [Courtesy NASA]

It was open, maybe, five seconds. We wanted to minimize that time as much as possible, because that was a break in containment."

It took the astronauts about five minutes to get into their BIGs. The hatch opened again, and Hatleberg assisted Aldrin into the raft that was fastened to the flotation collar, then immediately closed the hatch. Before Aldrin was allowed to sit down, Hatleberg sprayed and scrubbed the moonwalker's BIG head to toe with sodium hyper chloride—"a very strong bleach solution," Hatleberg says—in hopes of killing any "moon germs" that might be lurking. As he was doing so, he heard Aldrin mumble something that came out "mmmmmmrph" through the BIG's bulky facemask.

"I couldn't understand him," Hatleberg remembers. "So I knew he wasn't going to understand me, either. But as a Navy-trained communicator, when you hear an indiscernible message, you say it back. So I said, 'mmmmmmrph' back to him. He just looked at me and finally took a seat in the raft." Hatleberg performed the same decontamination procedure on Collins, minus the mumbling.

Armstrong, the flight commander, was last to leave the spacecraft. Hatleberg assisted the first man to walk on the moon into the raft with his crewmates, and then he quickly tried to shut *Columbia's* hatch. No luck.

"They had forgotten to recycle it," Hatleberg says. "It was like trying to shut a door that has a deadbolt in the open position. It would close, but not all the way. It wouldn't lock. I thought, 'Uh-oh, here we go.' In all the practice sessions, this had never come up. And I honestly didn't know what to do. So I just leaned against the hatch and turned toward the astronauts."

Michael Collins got up and stumbled over to the scorched spacecraft in which he had spent nearly fifty-seven hours orbiting the moon, many of them alone while Armstrong and Aldrin were on the lunar surface. "I guess he could see my angst, and he knew immediately what was wrong," Hatleberg says. "He recycled it, slammed it shut, and that was the end of it." Hatleberg sprayed down Armstrong, and Armstrong returned the favor.

Finally, the astronauts were securely in the raft.

"There were three things I had to remember at that point," Hatleberg says. "I had to check the vents on the top of the command module and make sure they were closed, meaning the atmosphere inside the command module was contained. Second, I had to make sure the tape on the astronauts' air filters was pulled off so they could breathe. And third—and I still have a hard time believing this one—[I had] to make sure the inflated water wings were properly placed around their arms. I still laugh about it. You have to remember, these people had been in a weightless environment for an extended period of time. We had to make sure that if they fell out of the raft, they would float. But I've never been able to get that vision out of [my] mind—people who had just come back from the moon wearing those bodacious, orange water wings."

The astronauts were hoisted into the recovery helicopter in the same order they exited the spacecraft and flown to the deck of the *Hornet*, which had arrived just moments after splashdown. Once onboard, they went directly into the Mobile Quarantine Facility (MQF), a trailer-like contraption that would be their boring home for the next twenty-one days. (It turns out they were carrying no moon germs, and Neil Armstrong celebrated his thirty-ninth birthday in confinement.)

Meanwhile, Hatleberg was still at work. He bathed *Columbia* with betadine before it was lifted onto the *Hornet* and placed near the MQF. He took his tool bar and punctured the flotation collar and sea anchor, sending them to the bottom of the Pacific. He took off his BIG, rolled it up, tied it to the astronauts' raft, and then sent both to the ocean floor in the same manner.

It was over. All the hours of practice had paid off. "I think I was relieved more than anything," he says. "If there had been a lunar pathogen, and we had screwed up out there, I'm the one who would've been responsible."

But as Hatleberg flew back to the *Hornet*, his adrenaline still pumping, it dawned on him that he had just been in the presence of men who had traveled to the moon. It was the first time he had allowed himself to think that way.

"That's when I really realized the effect the training had had on me," he says. "You focus on your job at the exclusion of everything else. I had actually blocked out the fact that I was interacting with the first humans who had ever walked on another world. It showed me why NASA and Apollo were so successful. It was all about the job and the procedures and doing whatever needed to be done."

Hatleberg was instructed to go through a debriefing with NASA officials before talking with the media. "NASA wanted to know if everything went as planned," he says. "So I sat down for three or four minutes, told them there was no break in containment, other than the fact that the hatch didn't cycle properly. They said, as far as they could see from the TV cameras on board one of the helicopters, that everything had gone well. And that was it."

Hatleberg exited the debriefing without any attention. Not a single member of the media on board the *Hornet* asked to speak with him.

They were focused on the astronauts, who provided photo ops through a window at one end of the MQF and President Nixon's in-person welcome home to the space voyagers. In a time before twenty-four-hour cable news networks, Hatleberg was, again, just another sailor in the Pacific. He went back to his quarters.

"I didn't participate in this for any personal notoriety," Hatleberg says. "So it certainly didn't bother me."

That night, lying in his bunk and reliving the day's events, he thought of the morning in Galveston Bay when he went through the splashdown procedure with the Apollo 11 astronauts, about a month before liftoff. He knew their names, but not their faces.

"We went out on a barge, got there early—well before the sun came up," he says. "I was standing over against the gunnels, waiting for us to depart, when this gentleman walked over. We got to talking, and he said he worked for NASA. I told him how much I'd wanted to be an astronaut

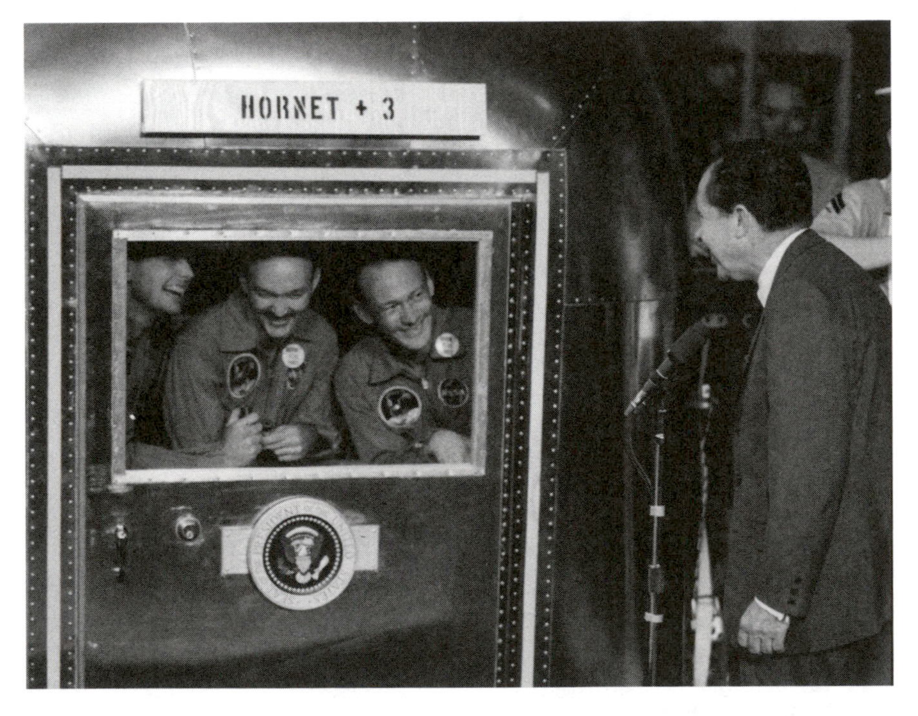

President Nixon greets the Apollo 11 crew through the window of the Mobile Quarantine Trailer, where the astronauts spent twenty-one days before being cleared of any "moon disease." [Courtesy NASA]

when I was younger, and he told me how much he would've liked to have been a frogman. He said, 'Well, I'd better get going.' We went on out and started practicing the procedure. The astronauts were wearing their BIGs. You couldn't see their faces very well through the visors. But I looked closely as I was washing one of them down, and I could see the eyes of the person behind the mask. It was Michael Collins, the guy I had talked with earlier that day! I couldn't believe it . . . Michael Collins!"

Hatleberg didn't see the Apollo 11 crew again until the day of the splashdown. He has visited with only one of them (Buzz Aldrin) since—at a thirtieth anniversary reunion aboard the *Hornet*, which is now a space and aviation museum in Alameda, California.

"I had a chance to say hello to him, chatted with him for a bit," Hatleberg says. "He was very nice and thanked me for what I did. I also asked him about what it was he might have been saying to me that day in the raft. He said, 'Heck, I don't remember. Maybe I was saying I wanted some fresh air.' So I never found out.

"These guys are private people, and I respect that. They have a lot of people wanting a piece of their time. Neil Armstrong is a very private person, and so is Michael Collins. I did write Mike Collins, and I relayed the story of how we had talked on the side of the barge that morning. He wrote back and said he remembered it. I mean, I hope to meet them all again some day, but if I don't, that's OK."

Apollo 11 was the only recovery Hatleberg ever worked. He received orders to attend the Naval Post Graduate School in Monterey in September 1969. He retired from the Navy in 1985 and has enjoyed success in the private sector, helping develop computer programs for businesses.

He and Sue have settled back in Chippewa Falls. They have two grown daughters: Jodie Hatleberg, who is a firefighter medic in San Diego, and Jamie Floesca, who resides in Auburn, Washington, and works in vocational rehabilitation for a hospital. Jamie is the mother of the Hatlebergs' two grandchildren.

Looking back at Apollo and his contribution, Hatleberg says: "There was a small army of people dedicated to the concept of getting those three men—and the ones who followed them—to the moon and back again safely. Everyone played their part and played it to the best of their ability.

"But it should never detract from the fact that on every one of those missions, there were three men who put their lives in harm's way. And we learned from Apollo 13 that it was a very real possibility that crews could get lost in space and never come back. Those are the real heroes.

They put their lives out there to advance mankind, and I would never want to detract from them at all.

"Apollo told us time and time again that the only limits to our endeavors is imagination. I'm very hopeful the space program will once again capture the imagination of the world one day. The whole world [was] unified [by] watching the most historic event of our time. We left this planet and came back. Amazing. How can a person *not* be touched by that?"

PART II

"We're Not the Soviets"

5 JULIAN SCHEER

NASA Public Affairs Chief

The television picture was fuzzy and colorless and absent of detail. It showed the outline of an astronaut, faceless behind a dark visor and moving stiffly in a bulky white spacesuit. In the background was a single shadow—that of the lunar module *Eagle*. It was cast onto a white-bright surface, and a slanting black horizon that fell quickly to the right—the camera was crooked.

Still, the broadcast, grainy as it was, recorded one of the most startling accomplishments in history: a human being's first steps on the moon. It allowed more than 700 million viewers around the world to share in the moment.

And the TV camera almost didn't make the trip.

Four months before Apollo 11 was to blast off, NASA scientists and engineers sneered at the idea of including a TV camera aboard *Eagle*. "They said the camera weighed too much—seven pounds, I think it was," says Julian Scheer, NASA's public affairs chief during the moon missions and a former newspaper reporter. "And weight was a critical issue, no question about it. But I insisted, 'You're going to have to take something else off. That camera is going to be on that spacecraft.'

"They said, 'No, no, you don't understand. It'll interfere with flight qualifications.' And they said, 'Our job is to get the astronauts to the moon and back safely, and bring a [soil] sample back—not to appear on television.' I even got a note from [astronaut chief] Deke Slayton that said, 'We're not performers, we're flyers.' He said there was no guarantee that they were even going to land on the moon on Apollo 11. They might get up there and decide not to do it. He said, 'There's really no reason for you to get excited about this.'

JULIAN SCHEER

Julian Scheer with his grandchildren (*left to right*): Noah Scheer, Adrienne von Shultress, Ginny Scheer, Richard Gerhardt. Taken in summer 2000 at Bald Head Island, North Carolina. [Courtesy Sue Scheer]

"They could never see the big picture. But they weren't landing on the moon without that [TV] camera on board. I was going to make sure of that. One thing I kept emphasizing was, 'We're not the Soviets. Let's do this thing the American way.'"

Newspapers were Scheer's first love.

"Everybody in my family grabbed the Richmond [Virginia] *Times-Dispatch* in the morning, and I got whatever was left over," he recalls. "Usually, that was the business section or the sports section. After reading it for a while, I really wanted to become a sportswriter."

In 1939, when he was thirteen, Scheer applied for a part-time job with a weekly newspaper chain in his hometown of Richmond. He worked there after school until 9 p.m. every weekday and then covered football games on weekends during the fall. He learned to meet tight deadlines, filing 150-word game summaries from press boxes via Western Union. He also began to look for things to write that went beyond the obvious.

After serving a stint in the Merchant Marines during World War II and graduating with a degree in English from the University of North Carolina in 1950, Scheer joined the *Charlotte News* as a sportswriter. "I thought, and still do, that it was the best paper in the country at that time," he says.

He spent about a year in sports, shifted to the political beat, then became the paper's featured columnist. Many of his articles focused on the civil rights movement.

In 1956, Scheer received an invitation that would change his life. "I had a friend, Nelson Benton, who was later a CBS correspondent. We'd been friends in college, and he was working at a television station

Julian Scheer (*center*) with President Kennedy. Scheer was a reporter with the *Charlotte News* at the time. This was taken during Kennedy's presidential campaign. It is a great photo in that it shows the person who challenged America to go to the moon (Kennedy) and the man who made sure Americans got to see the first steps on television. [Courtesy Sue Scheer]

in Charlotte [during] the same time I was there. On weekends, he flew with the Air National Guard. One day he came up to me and said that a colonel was going to take some people down to Cape Canaveral and asked if I wanted to come along. I had nothing else to do."

The Cape, an Air Force base at the time, was where the Department of Defense tested multi-stage rockets. Seeing firsthand what was going on made Scheer realize the United States' commitment to space exploration was real, especially as the Cold War with Russia escalated. After his initial trip to Cape Canaveral, he "became really interested in space" and "fascinated by it all."

In 1958, the year after Russia had launched unmanned satellites Sputnik I and Sputnik II into earth orbit, NASA was formed to lead America's charge into space. "That's when I had to make a decision," Scheer says. "I knew the two big stories of the century were going to be civil rights and the space program. I was doing a lot of freelancing for the *New York Times* and *Newsweek and Newsday*, all on civil rights. And I knew my paper wasn't going to pay for trips down to Cape Canaveral because no one believed that anything was going to come of the space program."

Always one to take the path less traveled, Scheer chose to turn his reporting efforts toward space. He paid his own travel expenses to the Cape several times, getting to know the strategy of the program and the people running it.

During NASA's first year of operation, Scheer co-wrote a book with rocket engineer Theodore J. Gordon, *First into Outer Space*. "It was a bestseller for a while because nobody had really done anything on space," he says. "But in those days, everything had to go through the Pentagon for [security] review, and they took big chunks out of it. Really gutted it."

Finally trusting the veteran reporter's instincts that the missions were worth covering, the *Charlotte News* dispatched Scheer to the Cape to write about the Mercury flights, beginning with the United States' first space voyage, a fifteen-minute suborbital flight in May 1961 made by Alan Shepard.

As interesting as the space beat was, it soon became frustrating for Scheer and other members of the print media, as they wanted to give their readers more insight than the NASA public affairs office was willing to share. They were constantly fed lame quotes from the astronauts and flight controllers. Only the TV networks were provided detailed flight plans. "We had maybe one press conference before a flight," Scheer says. "Four hundred to 500 reporters in a room trying to interview one astronaut. That's why we objected so much to the *Life* contract."

Life magazine had a signed agreement with the Original 7 astronauts, giving it exclusive rights to their personal stories. And the pay was sizeable at that time—about $25,000 annually per astronaut. So in July 1962, shortly after the fourth of six successful Mercury missions, Scheer quit the *News* to write a civil rights novel. He hadn't finished the first chapter before NASA administrator James Webb summoned him to Washington.

"He told me, 'You've covered this program, and we need your help,'" Scheer remembers. "NASA said it was an open program then, but it really wasn't. Webb felt, and so did I, that the American people ought to be able to witness everything. He told me to write a plan [for coordinating media coverage of the missions]. I spent about thirty days in Charlotte writing it, and I sent it to him. Wasn't long before he called and invited me back to Washington. When I walked in, he said, 'I accept your offer to go to work for me.'"

Both men laughed. Scheer grew serious and told Webb he wanted to spend the next year writing his novel.

"You can write that novel ten years from now," Webb said to him. "I want you to run this program just as you've outlined it. You'll work directly for me, and I'll do everything you say to do in this program until it fails in a way that I consider . . . a failure—then that will be your last day here. Do you have the guts to do that?"

Scheer thought for a moment, then plotted his future with one word: "Yeah."

He arrived at Cape Canaveral in time for the final Mercury mission, a two-day stay in earth orbit by Gordon Cooper, in May 1963. Scheer was stepping onto the turf of a legendary public relations official: John A. "Shorty" Powers, an Air Force Lt. Col. who had become known as the "Eighth Astronaut" and the "Voice of Mission Control." A showdown between the two was inevitable. When Scheer found out Powers had been the person catering to the TV networks, he was furious. He told Powers to "get the copy machine running" and give all members of the print media copies of the flight plan. Powers replied: "They don't need it."

"I decided to test the system and see which one of us was really in charge," Scheer says. "I fired him on the spot." Powers' official NASA bio says he "resigned" from his position in 1963 and retired from the Air Force a year later. For the first time in the brief history of U.S. manned spaceflight, print reporters were given the same press releases and updates as the TV networks. Webb never questioned Scheer about his decision to dismiss Powers.

"I really believe during that flight we began to build what we like to think was the most open program in the world," Scheer says. "The things Webb told me when he hired me were really interesting, especially looking back now. He said, as far as he could see, if we were lucky, we could get to the moon in ten years. Maybe. And if we were really lucky, we'd beat the Russians there. But in the meantime, the Russians were going to do everything first—the first multi-man spacecraft, the first spacewalk. He said, 'It's going to look like the Russians are technologically ahead of us. But the only real difference in the two programs is that theirs is military and closed, and ours is civilian and open. We can show them what a democratic society is all about.'

"I often say to myself that if the Russians had been first, they would've issued a press release in Moscow saying, 'The Soviet Union announces today that it landed two men on the moon at such and such time.' And a week later you might have seen a picture. And that was the difference. You wouldn't have had the thousands of people at the launch. You wouldn't have had the millions around the world watching the liftoff and hearing the astronauts and seeing them after they had landed [on the moon]. Those are clear demonstrations of the two great powers in the world at that time and what their philosophies were.

"It made you proud to be an American—and to be first on the moon."

A lot of people inside NASA weren't thrilled with the open program idea. Scheer explains, "This was an agency made up of people who had come out of industry—and industry is not a very open society. Others had come out of the military. And the military, of course, had never liked talking to the press.

"They made [this] case: If we fail, people will be able to watch the tragedy. . . . If we keep showing everything, Russia will copy our technology and get ahead of us. I don't think there was ever a day in my nine-and-a-half years at NASA where I said, 'Here's what we're doing to do' without some technical person saying, 'No, we don't need to do that.' It was a constant, never-ending battle."

But Scheer could be a pain in the butt when he needed to be. He even found a way to have a little fun with *Life*.

"I had objected to the contract as a reporter, and now I found myself managing it," Scheer says, laughing. "Their deal was for the astronauts' personal stories. So I made it a rule that they weren't allowed to ask any technical questions. If one of the astronauts told his wife, 'I was scared shitless when the heat shield rattled,' I would step in and say, 'Sorry, but that's technical. You can't use that.' *Life* got really annoyed with me, but I just didn't believe anybody should have exclusive rights to the accomplishments and deeds of a government employee. It wasn't right. That contract should've never been signed.

"And it created one of the few problems the astronauts ever had among themselves. It was fine when it was just the Original 7. Then we brought in more astronauts. And the first group didn't like it that the new guys were going to get an equal share of the pot. Suddenly, the pie was being sliced up into tinier pieces, so their devotion to *Life* magazine really fell off."

Even NASA's most science-driven employees realized the first landing on the moon couldn't be all business. There had to be a certain amount of ceremony and reflection. And since the National Aeronautics and Space Act, signed in 1958 by President Dwight D. Eisenhower, declared America's activities in space to be "devoted to peaceful purposes for the benefit of all mankind," the achievement should be shared with the rest of the world.

But how? NASA's brass turned to Scheer for answers.

"I had the privilege," Scheer says, "of sitting in a room with my staff and saying, 'We've just landed on the moon. What do we do? What do we leave? What do we say?' It all sounds so simple now, but it wasn't. A lot of people in Congress wanted a United States flag planted on the moon, but that was touchy."

A United Nations treaty declared "outer space, including the moon and other celestial bodies, [was] not subject to national appropriation." America wasn't out to make the moon its fifty-first state, but the treaty had to be satisfied.

"I had to sit down with the head of our international program and write on a piece of paper, which was put away in a file, that said by erecting an American flag, we were not claiming the moon as U.S. territory," Scheer says. "We wanted that in writing so no one could ever suggest otherwise.

"Somebody else in Congress wanted a United Nations flag planted, instead. But, of course, every country wasn't a member of the United Nations. That got to be fairly controversial, so we decided against it. But whatever we were going to do had to be worked around the astronauts' exploration duties. They had a lot of things to do in their two hours or so out on the surface."

Scheer submitted his suggestions to NASA officials and mission planners. To his surprise and pleasure, Neil Armstrong and Buzz Aldrin carried out each of them that summer evening in the Sea of Tranquility. They put up an American flag using a collapsible flagpole designed by NASA engineers. President Nixon spoke to Armstrong and Aldrin for two

Julian, as a reporter, with President Eisenhower. [Courtesy Sue Scheer]

minutes via phone from the White House's Oval Office. Nixon told the voyagers, "For one priceless moment in the whole history of man, all the people on this earth are truly one." Armstrong read from a plaque that was left on the lunar module's descent stage: "Here men from the planet earth first set foot upon the moon, July 1969 A.D. We came in peace for all mankind."

Armstrong and Aldrin spent the rest of their time taking pictures, testing their maneuverability in the moon's one-sixth gravity, setting up gadgets to measure seismic activity, and gathering 47.7 pounds of soil and rocks.

Prior to launch, Scheer reminded mission planners of one other procedure that had to be addressed: What if the astronauts were left stranded on the moon or died trying to get there?

"I still have in my files the contingency plans—what the president would say, what the head of NASA and the secretary of defense would say. Who would notify the next of kin," Scheer says. "All those emergencies were written out. And I exercised the discretion of cutting off the final conversations, if the astronauts couldn't get off the moon. I felt that if a crewmember wanted to say farewell to his family, he should have the privacy to do that. I knew there would've been a great clamor to release those tapes, and I knew that forever more I would've been criticized for not doing so."

He was used to it. Several newspapers ripped Scheer for withholding information immediately following the cockpit fire that killed astronauts Gus Grissom, Ed White, and Roger Chaffee on January 27, 1967, during a simulation for what was supposed to be the initial Apollo flight.

"When I found out what had happened, I put out a statement saying that we had had a fire during a test, named the crew, and said the accident involved a fatality," Scheer recalls. "I didn't say fatalities because we couldn't find Gus Grissom's wife. She was out shopping, and I could just imagine her listening on the car radio and hearing that her husband had been killed. I simply made the decision that the world didn't have to know before she did. So what if it took another half-hour or so? I thought a certain amount of humanity was due there, and I still think I made the right decision. We located Gus' wife within forty-five minutes, then released the full story.

"After all those years working in the newspaper business and working with reporters, it was the first time I questioned how much the public had the right to know."

Scheer was friends with all three astronauts, especially Grissom. "Gus was one of the original astronauts, and we'd always sort of pal around together when we went to the Cape. He had a '62 Corvette, and while he was out doing training, he'd let me tool around in it."

In fact, Scheer got to know most of the astronauts and their individual quirks. One of his favorites was John Glenn because he "was friendly and accessible." Scheer says reporters considered petitioning NASA to make Glenn the first American in space instead of Shepard. "We didn't know that they were saving him for the first orbital flight," he says.

After Apollo 11, Scheer accompanied Armstrong, Aldrin, and Michael Collins on a goodwill tour of thirty-seven countries in forty-two days to let the world see these men who had traveled to another world. Collins was loose, articulate, and seemed to enjoy the attention. Aldrin, in Scheer's words, "was a strange personality to deal with." Scheer won't elaborate.

And then there was Armstrong, a civilian from Wapakoneta, Ohio (population 6,756 in 1969), who Scheer believes was the perfect choice to be the first to walk on the moon.

"I always wanted a non-military person to be the first because it was a non-military program," he says. "It just seemed sensible that we'd rather have a Mr. Somebody be the first, rather than a major or captain or colonel. I think that did have something to do with Neil being selected as the first. But if that flight had been delayed or hadn't landed, then it would've been the next crew's commander, [Capt. Pete Conrad, United States Navy]. And that would've been terrific. Pete was a great individual."

But there was something about Armstrong, something that just fit the role.

"If you sat down and had a beer with him, you'd say he's a regular guy, without ego, very accomplished, very modest, not impressed with himself or what he did," Scheer says. "But he was very serious about his work, a highly-skilled test pilot, and very, very calm under pressure."

Scheer defends Armstrong's decision to live quietly and privately in recent years, rarely granting interviews or signing autographs: "To some that may appear negative. But that's because he's not a showboat. And I think a lot of his reticence to do certain things is because he doesn't think he's particularly important. Take it from me, he's a great guy."

Many of the astronauts felt the same way about Scheer. Hugh Norton, a longtime friend of Scheer's, developed Grandfather Mountain National Park in Linville, North Carolina. As a joke, Norton named one of the peaks Scheer's Bluff—complete with a sign. Scheer and his wife Sue often would take their children and grandchildren up the mountain to have their picture taken together.

To Scheer's surprise, he received a photograph in the mail of Apollo 8 astronaut Frank Borman standing at Scheer's Bluff. Borman wrote: "Julian, this is the first time I've called your bluff. We've been through a lot together and I've always valued your advice. Hang tough. Good luck and many years of happiness to a true friend."

Scheer never lost his love of writing. During the Gemini program, in 1965, he found time to pen a children's book, *Rain Makes Applesauce*, which was illustrated by Marvin Bileck. It remains in print today and was a runner-up for the Caldecott Award, presented annually, by the Association for Library Service to Children, to the author of the best American picture book. He released another in 1968, *Upside Down Day*, and two more during retirement: *A Thanksgiving Turkey* and *Light of the Captured Moon*.

Julian Weisel Scheer died of a heart attack shortly after the release of those books in 2001. He was seventy-five and was survived by Sue, four children, and four grandchildren.

Scheer had been correct in his assessment that civil rights and the space program would be the top stories of the twentieth century. And the 1960s, filled with war and protests and assassinations, were fertile ground for reporters.

During an interview one year before his death, Scheer said he had watched each event, whether it was the Rev. Martin Luther King leading a march in Selma, Alabama, or Ed White performing the first American spacewalk on Gemini 4, and wondered what news angle he would've taken had he been covering the story.

"In my mind, in my psyche, I was always writing," he said. "It never left me. I always got a charge out of seeing my byline in the paper, and I don't know why." He even missed battling deadlines, which really surprised him. And when it came to coverage of the early moon missions, Scheer paid close attention to the newspaper reporters and columnists. He mentally graded each one, looking for accuracy, originality and story development.

Of all the major newspapers, Scheer said the *New York Times* did the best job of capturing the significance of Apollo 11. He especially liked its headline on the morning of July 21, 1969: Men Walk On Moon.

Genius hidden in simplicity. Four words forever capturing history, as did the seven-pound, black-and-white TV camera that flew aboard *Eagle*.

6 JOSEPH LAITIN

The Reading of Genesis on Apollo 8

Joseph Laitin, a Hollywood freelance reporter in 1958, had spent a long, tough day putting together a ninety-second piece on the renowned French mime Marcel Marceau for his CBS radio entertainment program.

"Ever try doing an interview with a guy who pantomimes for a living?" Laitin says, laughing now. "It ain't easy."

Laitin had accepted an invitation to a dinner party that evening. "But I got home, and I was hot and really exhausted and didn't feel much like going out. So I called the hostess and said, 'Look, I feel awful. Would you mind if I took a pass tonight?' She got furious! So I said, 'OK, OK, I'll take a shower and be right over.'"

Because he dealt with movie stars on a daily basis, Laitin was accustomed to being in the presence of beauty. But when he walked to his assigned spot at the dinner table, Laitin became mesmerized by the young woman seated next to him. "She was more than beautiful," he says. "She was classy. Glamorous. I'd never seen a woman who would even compare to her."

She introduced herself as Christine, a ballerina from Paris who was passing through Los Angeles. She had married an American, she told Laitin, but was going through a divorce. They talked throughout dinner. At the end of the evening, Laitin—a forty-four-year-old bachelor—asked the twenty-eight-year-old dancer to have dinner with him again the next night. She accepted.

Conversation came easy to them, as if they had known each other for years. "After dinner I said, 'Would you be interested in the proposal of marriage?'" Laitin says. "She answered, 'Thirty seconds after I get a divorce.'"

Laitin chuckles. "Don't ask me why I did that. I had never been one of those people who believed in love at first sight. But it happened to me."

Their chance meeting and subsequent marriage in 1959 would touch the world in 1968 as Apollo 8 circled the moon on Christmas Eve night. Joe Laitin, with a valuable assist from Christine, came up with the idea for three American astronauts—the first humans unbound by earth's influence—to read the opening ten verses of Genesis during a live television broadcast to a stunned earth.

It not only was one of the defining marks of Apollo but among the most memorable moments in television history.

"I'll tell you what it was," says Jim Lovell, one of the Apollo 8 astronauts who took part in the reading. "It was absolutely perfect."

JOSEPH LAITIN

Joseph Laitin (*left*) and friend Simon Bourgin met during World War II. At Bourgin's request, Laitin helped come up with the reading of Genesis on Apollo 8, the first journey into lunar orbit. [Courtesy Simon Bourgin]

Joe Laitin grew up in Brooklyn and became well known for his news coverage of World War II. He began the war as a stateside correspondent for United Press International wire service, writing primarily about the war's economic impact on the United States.

"But I had always wanted to go to the war front and write from there," Laitin says. "Reuters, the British News Agency, asked if I'd like to join them. I said, 'Yes, as long as you send me to the front lines.' But they basically wanted me to do the same thing I was doing for United Press, so I turned them down. A month later they called back and said, 'Get fitted for a uniform.' I was off to the Pacific."

He arrived just in time to cover the United States' atomic bombings of Nagasaki and Hiroshima. He even finagled his way onto a reconnaissance aircraft to view the damage to the Japanese cities.

"Before I left for the Pacific, the Washington correspondents were briefed by the officer of censorship," Laitin recalls. "All we were told was 'Don't write any stories containing the words "nuclear" or "atomic"—not even in a comic strip.' So I knew something was up and began reading up on it the best I could. I was at General [Douglas] McArthur's office when the first one was dropped. Nobody knew what the hell it was, so I suddenly became the authority."

Reuters wanted Laitin wherever the major stories were breaking. He was aboard the USS *Missouri* at Tokyo Bay and wrote about the Japanese surrender. From there, he was off to Manila to cover the war crimes trials, and then to Germany for the Nuremberg trials. It was hectic but fulfilling. But eventually he tired of the travel and decided to try his luck in Hollywood as a freelance journalist, covering the movie stars.

"I tried breaking into the TV business," Laitin says, "and I did one or two small things. But that was a very closely knit operation in the old days. If you didn't know somebody, you were out of luck. And writing is what I did best anyway."

Laitin began freelancing for *Collier's* and the *Saturday Evening Post.* He also produced entertainment packages for CBS and ABC radio and wrote stories for a few newspapers. He was a well-traveled bachelor who made good money in a town of endless parties and women who were impressed by someone who hung out with the stars. It was a good gig.

In Hollywood, he ran across a buddy from the war, Simon Bourgin (pronounced BOR-gen), a native of Ely, Minnesota, and the former director of the Office of War Information. He had moved to Los Angeles as head of *Newsweek* magazine's West Coast bureau. Bourgin also was a bachelor who enjoyed the Hollywood dating scene.

"We'd throw parties together—he'd do the cooking; I'd handle the invitations," Bourgin says. "When I had to be out of the office, I'd get Joe to cover for me. He was very reliable and a really good writer. He did some pieces for *Newsweek*."

The partying ended, and Bourgin saw much less of his friend, when Laitin married Christine. "He treated her like a goddess," Bourgin says.

Laitin had lived an interesting life, but so had Christine. "She was a prima ballerina in Paris during the German occupation," Laitin says. "Because she was in the ballet, she could stay out past [the German-imposed] curfew, and she was authorized to have a bicycle to go back and forth to rehearsals. What the Germans didn't know was that she was helping run guns for the French underground.

"At a performance for German authorities, the chief German civilian representative of Paris turned to an officer and nodded toward Christine. He said, 'Get her name and address, send her a dozen roses, and find out if she'll have dinner with me tomorrow night.' Christine said she was terrified the Germans were coming to her house, because her mother was hiding a Jewish woman there that the Gestapo was looking for. When the officer arrived at the door, her mother said, 'She's always lied about her age. She's only fourteen years old, you know.' Of course, she was much older. But the Germans were proper, if nothing else, so that scared them away."

Laitin pauses, then says, "I keep going back to the fact of how beautiful Christine was, but it's true. She had blue eyes. . . . She was one of those Parisian women who had class and dressed beautifully. Whenever she walked into a room, everyone stopped talking.

"But she was an intellectual, too. She used to read a book a day; then [she] could discuss it with you in detail, which I couldn't. She was well versed in Shakespeare and the arts. Just a very unique woman."

Around 1961, Christine and Joe had a son, David, who at only five months old died of sudden infant death syndrome, now known as SIDS. The cause remains a mystery to doctors and researchers. "They used to call them crib deaths, and it certainly brought a lot of guilt in those days," Laitin says. "She had two children from her previous marriage [both of whom Laitin eventually adopted], but we really wanted to have a child together. It was rough. Finally, I had her go to a psychiatrist. He told her, 'You're wasting your money coming to see me. Nothing will take care of this problem except having another child.' But it had been a rough pregnancy, and the doctor had told her not to have another child."

They decided to try and adopt—"something I thought I was entirely too old for at the age of fifty." While the Laitins were negotiating an adoption, Christine became pregnant again. Another son, Peter, was born with no complications.

"Life was going well. Christine was taking care of the three children and doing some interior design work. But the pain from losing David was still there," Laitin says. "We had a house in the hills of Hollywood, and we decided it might be best to move somewhere new. Christine knew I had worked in Washington, so she suggested I look into maybe getting a job there.

"My horizons had broadened. So had my income. So covering something like the Department of Agriculture didn't have a lot of luster to me. But I decided to go and see what was available."

Once again, Laitin and Bourgin were about to cross paths.

By 1963, Bourgin was already in Washington, having taken a job with the U.S. Information Agency. His brother, Frank, was also in Washington, working with the Office of Emergency Management.

"Somehow, we found out Joe was wanting to come to work in Washington," Bourgin says. "I was talking to Frank and he said, 'Elmer Staats [head of the Bureau of Budget] is looking for somebody to handle pubic affairs. Joe would be perfect for that.' So I put Joe in touch with him."

Laitin was so sure he didn't want the job, he booked a 5 p.m. flight back to Los Angeles before going to the interview.

"[Staats] told me 'You're just the guy I've been looking for,' and I had just one question for him—why?" Laitin says. "I told him, 'I was a newspaperman years ago, covered economics and all that. But the last ten years I've sort of specialized, writing about sex and glamour.' He held up this big stack of papers with all the budget information in it, dropped it on the floor and said, 'That's *exactly* what this damn thing needs!' So I took the job, then called Christine and told her to put the house on the market."

Christine and the children joined Laitin in Washington a few months later—just about the time President Lyndon Johnson offered Laitin a job.

"I must say it was kind of impressive; I wasn't used to getting a call from the president," Laitin says. "I went over to the White House to talk with him, and while I was waiting outside the Oval Office, his top speech writer walked up to me and said, 'If you're Joe Laitin, the president is waiting for you. Go on in.'

"So I go in, and he's not there. I hear the dribbling of water, and I look back, and there he is peeing in his private bathroom with the door open. He yelled out 'I'll be with you in a minute!' He finally comes out and says, 'Listen, my press secretary tells me he's got these shitfaces calling him all hours of the night.' That was the only way I ever heard him refer to the media—shitfaces. He said, 'We need you over here to take those calls.' So that's how I wound up in the White House."

He served as assistant press secretary to Johnson and worked in the administrations of three other presidents—Richard Nixon, Gerald Ford, and Jimmy Carter.

In the summer of 1968, the Apollo program was struggling to find its feet. A fire on the launch pad in January 1967, during what was thought to be a routine test, had claimed the lives of Gus Grissom, Ed White, and Roger Chaffee—the crew of what was supposed to have been Apollo 1. Some members of Congress, particularly Sen. Walter Mondale, had been waiting for any excuse to shut down the manned space program. Mondale, a Minnesota democrat, deemed Apollo a waste of taxpayers' money and pushed to end government funding.

During the investigation of the fire, Frank Borman was asked to help restructure the Apollo command module to make it safer. Borman, who had commanded Gemini 7, was one of the most respected leaders on the astronaut roster. After visiting with the engineers, who designed every component of the vehicle, and studying its intricate systems, he came away more determined than ever to prove the naysayers wrong about NASA, the command module, and all the people who worked on the space program.

Borman had been named commander of Apollo 8. His crew consisted of Jim Lovell, who had flown with him on Gemini 7, and rookie Bill Anders. The mission was scheduled to test the LM, a particularly complicated spacecraft that eventually would be used for the actual landing on the moon. But there were numerous problems in its development. And since Apollo 7 was scheduled to test the command module's systems during an eleven-day, earth-orbit flight in October 1968, mission planners were wondering what to do with Apollo 8. If they waited on the lunar module, it could push the launch all the way to February. Even after the fire, NASA's goal was the same: Land Americans on the moon by the end of the decade, bring them back safely, and do so before the Russians. And the pressure was mounting. While struggling to develop a rocket capable of matching their dreams, the Soviets were desperately pushing to send a cosmonaut around the moon before the end of the year.

George Low, one of the engineers who helped develop the new and improved command module, pitched an idea: Instead of sending Apollo 8 to orbit the earth, or waiting until February for the LM to be ready, why not send Borman, Lovell, and Anders to orbit the moon?

It took some selling to convince NASA brass. After all, Apollo 7 hadn't even flown yet. Traveling to the moon held so many unknowns. Another catastrophe would almost certainly be the end of Apollo. But the more NASA thought about it, the more sense it made. It might just be the morale boost the program needed.

When asked if he was willing to take his crew to the moon—provided Apollo 7 was a success—Borman immediately answered yes. Anything for the program. Anything to beat the Russians. So on October 22, as Apollo 7 descended into the Pacific Ocean having completed a near flawless mission, Apollo 8's destination became clear.

Not even the most optimistic people in the space program dreamed a moon mission would be attempted this soon. NASA had flown just seventeen manned missions, beginning in 1961: six Mercury flights, ten Gemini, and one Apollo.

At a time when color television was still a big deal, what Borman, Lovell, and Anders were hoping to accomplish was almost beyond comprehension: Leave our planet's gravitational force by riding a rocket with 7.5 million pounds of thrust; view the total earth suspended in space the way school kids glance at a globe; fly within sixty miles of that mystical, romantic object in the sky that controls our tides and, some say, our emotions; soar behind the moon and out of communication with another human to see mountains and craters no eyes had ever witnessed; then return to earth by barreling through a narrow alley in the atmosphere at 24,000 miles per hour before splashing down in the Pacific.

Science fiction was about to become reality, and the drama only increased when NASA realized Apollo 8 would reach lunar orbit on Christmas Eve. A primetime telecast live from the moon was planned. NASA public affairs urged Borman to come up with something that would capture the moment.

A few weeks before launch, Borman phoned Si Bourgin.

Bourgin was still working with the USIA, now as its science advisor. "I was really a liaison, in charge of the foreign press during Gemini and for the upcoming Apollo program," Bourgin says. "I came to know the astronauts really well. After they flew, the United States would send the astronauts to different parts of the world as goodwill ambassadors. I would advance the trips and then accompany them. We went to a lot of places—South America, Korea, Japan."

Bourgin became especially close to Borman after Gemini 7. "Frank turned to me a number of times, when he was about to be interviewed, and he would ask me how to answer certain questions," Bourgin says.

"He came to trust me, and I dined at his and [his wife] Susan's home on a number of occasions."

Borman had no clue what to say to the world on Christmas Eve; he asked Bourgin for help.

"I tried writing something out," Bourgin says, "but it just didn't sound right. So I called Joe Laitin, told him what was going on, and asked if he could help come up with something. Then I said, 'By the way, Frank needs it in twenty-four hours.'"

Laitin was used to pounding out stories in no time and assured Bourgin it would be "a piece of cake." Laitin quickly found out he was wrong.

"I had someone do a drawing for me of what they *thought* the earth and moon would look like to the astronauts that night. Hell, nobody knew!" Laitin says. "It was a black-and-white rendering. The moon was large, and the earth was small. Looking now at pictures taken by the astronauts, the artist had darn near everything right.

"So after dinner, we put the kids to bed. Christine went upstairs to read. I propped up my portable typewriter on the kitchen table, and I put this drawing in front of me. I looked at the sketch of the earth, which was about the size of a tennis ball. And I thought, 'Everything I love is on that tennis ball. What would I feel like saying to it on Christmas Eve?'

"Pretty soon, I saw the problem. Everything I wrote was about peace on earth, and here we were in a war in Vietnam! Writing about peace on earth at that time would have made us all look stupid. By now, the floor was littered with balls of paper, and I was beginning to get a little frustrated, because I considered myself a professional.

"So I decided to back up a little bit. I asked myself again, 'What am I looking for?' And the answer was something that would go with Wurlitzer organ music. Something Biblical. So I finally went upstairs and got my Bible. Now I felt like I was onto something. They would be orbiting the earth on Christmas Eve, the day before Christ's birthday. I turned to the New Testament and the story of Christmas, . . . but I soon realized this wasn't what I was looking for either. So, here I [was] at 3:30 in the morning, crumpled paper all over the floor, reading the New Testament, and *really* becoming frustrated.

"Suddenly, my wife [came] downstairs [and saw] her husband reading the Bible in the middle of the night. She said, 'My goodness, Joe, what have you done?' I explained to her what I was trying to do, and she said, 'Well, if you're looking for that kind of language, you're in the wrong part of the Bible. You need to go to the Old Testament; that's the kind of writing you're looking for.' So I got very irritated with her. I said, 'It's 3:30 in the morning. I wouldn't even know where to begin.' She said,

'Why not begin at the beginning?' So I flipped over to the first chapter of Genesis, read the first verse, and said, 'Christine, here it is!'"

Laitin sent his typed message to Bourgin the next morning with a cover letter that read: "Si, I think this is what I would feel like saying to the earth. But tell Borman if he and his crew don't feel the same way, throw it away." "Then I forgot about it. I didn't realize I was writing what would become a footnote in history."

Laitin laughs. "Let me correct that. I didn't *write* Genesis, okay? But I made some other notes and suggestions."

Bourgin sent it on to Borman. After a few days without hearing back, Bourgin phoned the astronaut. Bourgin recalls, "Frank said, 'Oh, I meant to call you. I just took what you sent, scissored it out, had it put on fireproof paper, and stuck it in the back of the flight plan.' That's the last I thought about it."

The first descriptive words spoken from lunar orbit were haunting. They came from Jim Lovell, a forty-year-old, former Navy test pilot, as he looked at the moon from sixty-nine miles away: "The moon is essentially gray. No color. Looks like a plaster of Paris . . . or sort of a grayish beach sand. We can see quite a bit of detail."

It was approximately 4 a.m. Houston time, and the world hung on every word as Apollo 8 made the first of its ten orbits. The flight plan was chock-full of landmark tracking and photography assignments. Mission planners wanted vertical and horizontal shots of several possible landing sites for future flights, and also 16mm film of the areas. Those chores, plus getting enough sleep to be fresh for their critical Trans Earth Injection burn scheduled for shortly after midnight, took up most of their day.

But the highlight of the mission—of *any* mission, some still argue—began at 8:11 p.m. when Borman, Lovell, and Anders turned on the black-and-white TV camera. They showed their view of the moon to approximately half a billion viewers in sixty-four countries, all located on the beautiful blue ball that kept dipping in and out of the command module's windows.

"This is Apollo 8 coming to you live from the moon," Borman said, opening the program. "Bill Anders, Jim Lovell, and myself have spent the day before Christmas up here, doing experiments, taking pictures, and firing our spacecraft engines to maneuver around. What we'll do now is follow the trail that we've been following all day."

The TV picture quality was amateurish by today's standards but clear enough to make out craters and mountains. It was breathtaking.

Apollo 8 crew William Anders, James Lovell, and Frank Borman became the first humans to ever see the earth as a whole. [Courtesy NASA]

Si Bourgin and his wife, Mariada, awaiting a flight to Washington, watched from an airport lounge in Houston. Approximately forty people sat with them in a silence that bordered on reverence. In Bethesda, Maryland, Joe and Christine Laitin were curled up on the couch, words of humans 250,000 miles from earth filling their living room.

Borman said, "[The moon] is a different thing to each one of us." He went on to call it "a vast, lonely, forbidding type of existence or expanse of nothing." Borman asked Lovell to share his thoughts. "The vast loneliness up here at the moon is awe-inspiring, and it makes you realize what you have back there on earth," Lovell said. "The earth from here is a grand oasis in the big vastness of space." Anders added, "The thing that impressed me the most was the lunar sunrises and sunsets. These in particular bring out the stark nature of the terrain."

The telecast was in its final moments now. With Anders holding the sheet of paper Joe Laitin had typed, and Lovell pointing a small flashlight toward the words that were at best dimly lit, Anders, a thirty-five-year-old Naval Academy graduate, said, "We are now approaching, uh, lunar sunrise. And, uh, for all the people back on earth, the crew of Apollo 8 has a message that we would like to send to you."

As the TV camera continued to focus on the moon passing by the window, Anders began reading. "*In the beginning, God created the heaven and the earth; and the earth was without form and void, and darkness was upon the face of the deep, and the spirit of God moved upon the face of the waters. And God said, Let there be light. And there was light. And God saw the light, that it was good. And God divided the light from the darkness.*"

Anders passed the paper to Lovell, and Borman held the flashlight for him. Lovell read: "*And God called the light Day, and the darkness he called Night. And the evening and the morning were the first day. And God said, Let there be a firmament in the midst of the waters, and let it divide the waters from the waters. And God made the firmament, and divided the waters, which were under the firmament, from the waters which were above the firmament. And it was so. And God called the firmament Heaven. And the evening and the morning were the second day.*"

Lovell handed the paper to Borman. After an eight-second pause, Borman read: "*And God said, Let the waters under the heavens be gathered together unto one place, and let the dry land appear. And it was so. And God called the dry land earth. And the gathering together of the waters called he seas; and God saw that it was good.*"

Borman then added a closing remark: "And from the crew of Apollo 8, we close with goodnight, good luck, a Merry Christmas. And God bless all of you, all of you on the good earth."

The screen went blank. Communication from the spacecraft went silent. And the millions of viewers on earth were left to ponder what they had just witnessed.

One thing was certain: Our world would never be the same.

In Houston, Joe Laitin looked at Christine in disbelief.

"That was my script!" he said.

"Don't you mean *our* script?" Christine laughed.

Their phone rang. It was Si Bourgin calling from a payphone in the Houston airport. "Joe, did you just hear what I heard?" he asked incredulously.

The beauty of earthrise caught the Apollo 8 astronauts off guard, but William Anders managed to take this photo with a long-range lens. [Courtesy NASA]

Immediately after the reading, reporters wanted to know the story behind it. NASA officials had no answers—they were caught off guard just like everyone else.

Soon after Borman returned to earth and went through his official debriefings, he had dinner with Si and Mariada Bourgin.

"When I sent Frank the note, I never told him that the idea had come from Joe," Bourgin says. "It wasn't to take credit. I just knew he had a lot on his mind, and I decided to tell him when he got back. So after dinner, I told him, and I asked him to call Joe, which he did."

Recalls Laitin: "Frank thanked me and said he was going to give me full credit when the round of press conferences began. And I said, 'Frank, don't do that. If you do, the Soviet Union will use it as propaganda, saying that the message wasn't truly from the astronauts but from some guy in the budget department.' He said, 'But I'm not going to lie when they ask me.' So I said, 'Well, I'm not asking you to do that. Just say that you received tens of thousands of suggestions—which I'm sure you did—and mine was one of them.' And that's the way he handled it."

A footnote about the reading: Laitin wrote the closing remarks too—"And from the crew of Apollo 8." Laitin also suggested turning off the TV camera immediately, as not to give mission control a chance to add anything to it. He had learned a thing or two about drama in Hollywood.

"Borman also told me that [when they were] about a day away from the moon, he took out the sheet of paper I'd sent, showed it to the other two crew members, and told them to go through it and let him know if they didn't feel comfortable with it," says Laitin. "The one thing Borman did do, which I thought was brilliant, was distribute the paragraphs among his crew."

Laitin is still amazed by one thing: "All three of them did the reading without a single slip-up. But that was sort of like the flight itself. So many things could've gone wrong. Sometimes you just luck out, that's all."

Bourgin wasn't surprised. "All the astronauts—every single one of them—were superior people. They couldn't have gone through the [selection] process and be chosen without being superior. They were highly trained, highly skilled people who kept their cool when times got tough. They could handle anything."

It had been a disturbing year in America. Robert Kennedy and Dr. Martin Luther King were assassinated. The war in Vietnam was escalating; so was the number of bodies being shipped home. But at least we had Apollo 8, which returned safely December 27.

"Looking back, two things really stand out about that mission and the reading of Genesis," Bourgin says. "First of all, the reading could *never* take place today. Liberal [and] right wing religious groups would be all over it. Our climate has changed so dramatically.

"And the other is that it's very difficult to explain to generations today what that mission and television broadcast meant. If you live long enough, things sort of lose their currency, after about twenty or thirty years, because the up-and-coming generation has no capacity to understand it.

"Just look at America today; people talking on cell phones as they're walking down the street, as they're riding in buses. There's so much noise, we can't hear ourselves anymore. That night in the airport lounge, as the astronauts read from Genesis, there wasn't a single word spoken. Not one. I'm not sure that would be the case today.

"And let's remember one other thing: When they did that reading, they still had to make the burn to get back home to earth. We didn't know for sure if they'd make it back or not. Thank goodness they did, but I never hear anybody even mention that part of it."

(Joe Laitin died of heart failure Jan. 19, 2002 at the age of 87 at his home in Bethesda, Maryland, about two years after the interviews with the author. Christine Laitin preceded her husband in death in 1997, following a lengthy illness.)

7 HUGH BROWN

Telemetry and Communications, Kennedy Space Center

They were out there: Russian submarines, sneaking around off the coast of Florida, trying to jam NASA's communications and tracking systems during Apollo launches, desperately hoping to sabotage America's race to the moon. Hugh Brown knows it the way a parent senses when his or her child is in danger. And enemy submarines aren't foreign objects to Brown, who flew reconnaissance missions in the 1950s along the east coast of the United States in the Air Force's RC-121 Constellation.

Other members of the Apollo launch team are convinced of it, too. Even the astronauts say they had heard rumors the Russians had tracking ships near the Cape during launches to collect data and study American technology.

"One of the things we had found during liftoff on a couple of the early Apollo missions was that when the astronauts would speak, their transmissions would cut out—not completely, but they were garbled at times," says Brown, a telemetry and communications specialist at the Kennedy Space Center (KSC) during Apollo and an expert in electronic countermeasures, having worked with the Air Force for four years and with International Telephone and Telegraph Corporation (ITT) for five. "With the intricate [communication] system we had, there was no reason that should be happening. . . . So the question was, what was causing it?

"Initially, we had no inkling. But we did find out, through our tracking systems, that there were Russian submarines in the area. They were in International Waters, so it wasn't like we could keep them out of there. But jamming our systems certainly would've been in their best interest,

HUGH BROWN

Hugh and his wife Ina Brown. Hugh Brown was one of the few African Americans who played a part in the Apollo program. [Courtesy Hugh Brown]

at the time, because they did not want us to succeed. And with the proper scanning equipment, it wouldn't have been that difficult to figure out our frequency and send out a signal to interrupt things. So we had to take countermeasures."

Brown volunteered for the assignment. He had come to enjoy unique challenges.

The son of a Methodist minister, Brown was one of the few African Americans involved in America's space program. NASA, along with the companies that performed contract work during Apollo, was a reflection of society's workforce in the late 1960s—mostly white, mostly male.

"It was pretty obvious every day when I went to work that there weren't many people the same color as me," he says. "To me, that was a platform [for] *having* to be successful. I wanted to make sure I achieved every objective that was put before me—for [ITT] and for NASA. Sure, it put extra pressure on me. And there were times when I heard what might be considered racial comments. But it wouldn't ruffle me to the point of changing my attitude or my approach to my job. I had to maintain my own integrity and my own dignity.

"I just wanted to do my job and help beat the Russians to the moon. Everything was riding on Apollo. It was critical to our country. Being the first to land on the moon came before everything. I think you had to be a part of it to really understand what I mean by that. The only thing I can equate it to, when we would watch those Saturn rockets soar toward space [or make] their way to the moon, was someone giving birth. It was that intense, that important."

Brown, born July 16, 1935, was the third youngest of eight children. He attended six different schools across Florida between the fourth and twelfth grades. "That's part of being the son of a Methodist preacher," he says. "They'd move Dad about every year-and-a-half. But it made me learn to adapt to new surroundings. Some people may look at it as a negative, but I think it made me wiser. I had to make new friends and leave old ones. But I found there was a basic characteristic in people everywhere I went—that we all want to live and survive and succeed."

No matter where their daddy happened to be preaching, all of Rev. Morris and Annie McAlpine Brown's children were active—singing in the choir, serving as ushers, reading Scripture to the congregation. They especially enjoyed decorating the church for Easter services, fetching Palmetto limbs from the woods to place around the altar, and helping their mother make a sign of crepe-paper flowers that would spell out the word, EASTER and hang in the back of the church.

In his free time, Brown was an entrepreneur. He would collect people's shoes on Saturday, take them home and polish them, then deliver them that night in time for Sunday service. He had a paper route. He raked yards.

When he was eleven, Brown helped clean an office building in Sarasota. One day the woman who usually operated the elevator didn't show up for work. Brown volunteered to take her place.

"They didn't think I could do it," he says, "because in those days you didn't have elevators where you just pushed a button, and it took you to a certain floor. It had a lever you had to operate to go from one floor to the other. But I was real inquisitive as a child, and I had watched this woman operate the elevator. They couldn't believe it when I got on there and showed them I could do it."

He laughs again, "I [was] always getting these looks that said, 'Kid, get away from here.' Growing up in Florida, I was always around the citrus crops, and I used to see these guys hauling five and six crates of fruit on a dolly, and I was intrigued by that. I bugged those men to death until they finally let me try it one day. And I did it."

Brown says his philosophy was "work, work, save, save." The only thing he can ever remember splurging on was a Schwinn bicycle that had every gadget imaginable. "I used that for my paper route and to get me to my other jobs," he says. "So it was sort of like an investment.

"Really, that time in my life set a platform for me. When I saw something I wanted to achieve or something I wanted to buy, I may not have been able to get it right then, but I would always say, 'One day.'

"I was always telling my parents that I was going to do this and that when I grew up. My mom would say, 'Son, you can do it.' And dad would say, 'Son, you've got to work hard.' Between them, there was always a balance. They were the pillars of my foundation."

Brown also earned good grades. He was particularly fond of his high school industrial arts class. He built coffee tables, end tables, and chairs. And he never needed a book to show him how.

After graduating in 1954 from Lake County Training School in Leesburg, Florida, he earned a full scholarship to Northwestern University in Evanston, Illinois. "When I visited the campus, it just blew me away," Brown recalls. "It was a great school, beautiful campus. But it was *huge*—a city unto itself. I just wasn't prepared for that after being raised in smaller towns around Florida. For some reason, I just didn't feel comfortable. So I decided to enter the Air Force instead."

Brown studied airborne communications, electronic countermeasures, and radio operations, at Keesler Air Force Base in Biloxi, Mississippi, and was then transferred to Okinawa for two years.

"The Korean War had just ended, and I was doing a lot countermeasures in the C-47 and C-119 [airplanes]," he says. "We were still flying support for our fighter aircraft in the area, and in order to assist, we used to fly those different aircraft into areas, specifically to jam the enemy's radar, so they couldn't see our fighters coming through. We were in a danger zone and had our tail shot off a few times. But we always managed to make it out of there."

His next stop was Otis Air Force Base, near Cape Cod, Massachusetts. Here, he flew the RC-121 Constellation along the east coast as part of the 960th Airborne's Early Warning and Control Squadron. "That was quite a plane," he says. "We'd have a crew of between fourteen and sixteen people doing communications, radar, and airborne surveillance." Brown was making a name for himself, earning Airman of the Month as a flying radioman.

But by the fall of 1958, Brown was ready to take what he learned in the military and apply it in the private sector. He enrolled at Central State University in Wilberforce, Ohio, and in three years, graduated with a math degree. He interviewed with ITT and was hired the same day.

In 1962, Brown became part of the ITT team in Paris that maintained the communication systems for the Supreme Headquarters Allied Powers Group, an integrated allied military command for the defense of the NATO nations of Europe. It was one of the most complex systems at the time, involving high-powered transmitters that scattered signals off the troposphere, the bottom layer of the atmosphere.

Brown learned amazing new technology nearly every day. He even got a lesson in politics, while working in Paris. French president Charles de Gaulle was unhappy with NATO's integrated military structure. He felt France didn't have enough clout in the group. France withdrew from NATO, and NATO had to find a new site for its central communications command post.

"That was an interesting time," Brown says. "Out of the blue, de Gaulle just told us to get our stuff and leave his country. Well, that was easier said than done. It meant we had to move everything [which] was centrally located in that one place. You're talking about one small area where all these communications signals were bundled up and directed back to antennas all over Europe. But again, it was a challenge, and I enjoyed it."

Supreme Headquarters moved out of France in July 1966 and relocated to Casteau, Belgium. ITT completed the system and had it operating on March 31, 1967.

Brown was ready for a change.

"Fortunately, I had a lot of opportunities to transfer back to the States," he says. "I looked at them all." One stood out: ITT had won a communications and instrumentation contract with NASA. Apollo was almost ready to fly.

"It couldn't have been a better fit for me," Brown says. "Not only did it allow me the chance to come home to Florida, where I grew up, but it also brought me back to something I had a high degree of interest [in] and passion for. Really, it was one of the proudest days of my life when I went to work at Kennedy Space Center."

In October 1967, Brown had his picture taken near Launch Pad 39A with the Apollo 7 spacecraft and its Saturn IVB rocket poised and ready for launch. Brown was wearing a short-sleeve white shirt, a tie, khaki pants, and a white hard hat. A bundle of file folders rested under his left arm.

When he showed the photograph to his parents, during a trip home after the flight, they kept looking down at the picture, then back up at their son. "They said, 'You're *really* working with the moon program?'"

Hugh Brown, shown here as a specialist with ITT, helped monitor for Russian subs hoping to jam communications and tracking during Apollo launches. [Courtesy Hugh Brown]

Brown laughs. "It was hard for them to grasp. They were so proud. *So* proud. I'll never forget the looks on their faces that day."

After arriving at Kennedy Space Center, Brown was schooled to be part of the telemetry ground station team—a critical part of launch operations. "We were involved in getting all the 'bits' that were being transmitted from the spacecraft and the launch vehicle," he explains, "and those bits were recorded on magnetic tape, one-inch wide. The telemetry ground station took all those bits and was able to [plot the] status of the spacecraft and the launch vehicle at all times during pre-launch and launch. We had access to every system, every component on the spacecraft, and the launch vehicle.

"I loved it. It not only allowed me the chance to speak to other individuals in the communications industry, but also with the folks in mission control [in Houston]."

Brown soon became a shift supervisor, then chief supervisor of the thirty-six-man team of engineers and technicians, who were versed in the language of magnetic tape and who specialized in different areas of the vehicle. He also helped coordinate tracking.

"We had antennas in place to track the vehicle during ascent from the pad and get real-time data fed back to the telemetry ground station. It was sort of like monitoring a person's heartbeat."

Once the spacecraft was in earth orbit, Brown's crew translated and fed the data that had been recorded during launch to mission control. Every component was evaluated before a "Go" was given for Translunar Injection.

For Apollo 11, Brown's attention was directed toward the Russian submarines. Everyone, including the Soviets, knew America was running out of time to meet President Kennedy's 1961 mandate to land a man on the moon and return him safely before the end of the decade. It seemed the United States was surely going to win the race between the two super powers, but it meant much to many to be able to do it in the 1960s.

"We still weren't sure where the Russians were technology wise," Brown says. "They'd had some failures in their space program, but at that point we could take nothing for granted. I remember [in 1957] when they launched [the unmanned satellite] Sputnik. We thought, as Americans, that we were at the apex of the technology arena. Personally, I thought we were leading in every application. But Sputnik not only got our attention, it [also] made us ask the question, 'Where *are* we technology wise?' It created an absolute passion in the technology community to step up and be equal to the task. We simply could not allow Russia to be technologically superior to us. It was a critical time."

Nor could NASA take any chances of Russia ruining Apollo 11's moment. And a jammed communications network was more than a nuisance—it was dangerous. Astronauts racing for space at nearly a mile per second just two minutes into the launch procedure needed to be able to talk with the people on the ground who were monitoring their spacecraft's systems.

"We decided, in order to make a clear determination of what was causing these garbled transmission, that we needed to put a specific type antenna on top of the VAB [Vehicle Assembly Building]. It was the tallest structure at KSC, about 530 feet up. We needed to be high, and we needed to be able to look in all directions. From that vantage point, if there was a signal trying to disrupt communications, we could pinpoint the location, stop the countdown, and take corrective actions."

Brown wouldn't say exactly what "corrective actions" meant. It might have been something as simple as changing radio frequencies. But this was an issue the United States took seriously.

"Understand, we could do nothing about the Russians being in International Waters," Brown says. "There are Russian subs out there today. But if they were trying to disrupt things, then that was another story."

The surveillance dish was huge—eighteen-feet in diameter. "We couldn't get it into the building's elevators to take it up to the top," Brown says. "We tried using a helicopter, but the winds coming off the Atlantic made it very risky. So we came up with the idea to cut it half, then into quarters, and then reassemble it on the roof."

Apollo 11 lifted off right on time, and just as the vehicle cleared the tower, commander Neil Armstrong's voice came in loud and clear from the command module's cockpit: "OK, we've got a roll program." No interference. No garbled voices. And the antenna Brown was monitoring remained silent.

He often wonders if the sight of that huge dish atop the VAB was simply enough to discourage the Soviets. Surely they knew what it was for.

Four days later, Armstrong and Buzz Aldrin walked the surface of the moon, while command module pilot Michael Collins kept watch from lunar orbit.

The United States had done it. And history would record that Hugh Brown—a dreamer who grew up in a modest household and an African American who in his own way gave hope to his race—took part in the greatest scientific and aviation achievement ever.

HUGH BROWN

Hugh Brown during his days as a communications specialist with the U.S. Air Force. [Courtesy Hugh Brown]

Even today, he is humbled by the magnitude of it all. "NASA not only helped spur technology, in terms of going to the moon and back, but [also] the spinoff from that was absolutely awesome for all of humanity," he says.

"NASA raised the bar in terms of technical competence. Now, it's struggling again. All your General Electrics and your General Motors . . . they're all kind of scrounging people away who during Apollo would've gone to work at NASA. Today, the young engineers and technology specialists are going for the exorbitant salaries and working on things like cell phones and laptops, instead of rockets and spaceships. All the top minds are over in private industry doing marvelous things. Somehow, NASA has got to be able to reinvent the wheel, so to speak, as far as attracting young minds and setting goals that create that passion again.

"There have been overtures about going back to the moon and then on to Mars, but that is going to take funding. And, unfortunately, the government, a lot of times, gives lip service to things and then doesn't put its money where its mouth is. NASA will suffer until Congress is willing to do that again. Who knows when that will be?"

Brown worked the remaining six Apollo missions and eventually left ITT in 1977 to start his own engineering and technical services company. "I gave up a vested pension plan and a great salary because I had a vision," he says.

Getting started was tougher than he ever imagined. "I couldn't get a loan," he says. "Not a single bank would step up and help me. There weren't a lot of African-American companies in the technology management business. I didn't own a construction company with all these tangible assets. All I had was my expertise. They told me if I got a contract, then they would talk to me."

Brown kept bidding on jobs. In 1978, Pan American Airways verbally agreed to a contract with him to maintain its computers and communications systems.

"But they said it would be sixty to ninety days before things could be finalized," says Brown, a husband and father of two young children at the time. "Meanwhile, I've got no way to pay my bills. So I went back to Kennedy Space Center and said, 'I need a job for ninety days. Can you help me?' They put me in logistics with the Rockwell Corporation [where I helped] secure tiles for the space shuttle."

Exactly ninety days after Brown went back to KSC, Pan Am called. The contract was his. Brown hired a staff and went to work. He also went back to the banks that had promised to do business with him, only to be turned down.

"I needed $16,000 to meet payroll that month," Brown says. "My wife [Ina] was on her way to deliver the payroll checks when I called her and told her the banks were still turning me down. She said, 'What do you want me to do with these checks?' I thought for a minute and said, 'Baby, hand 'em out.' She said, 'Are you sure?' I said, 'Yeah, hand 'em out.'"

This was a man who, as a child, made kites out of newspapers and tree limbs. He had helped put men on the moon. The one thing both experiences taught him was to simply believe.

"And I knew God wanted me to succeed," he says. "Everything I had ever needed had been given to me. I needed a job for ninety days; I got one. I needed a contract; I got a contract. So I just knew that somehow this thing would work out. I went strictly on my faith and instincts."

Brown phoned his attorney at 3 p.m.—just about the time his employees were receiving their paychecks. He put Brown in touch with SunTrust Bank in Cocoa, Florida, and at 6 p.m. the funds were in his account.

His company, BAMSI, grew to 2,200 employees in twelve states and achieved $90 million in sales for eighteen consecutive years. Brown retired in 1997 and phased out the company a year later.

Along the way, Brown became a board member of SunTrust Bank and chairman of the Federal Reserve Bank in Atlanta, often swapping ideas with economic guru Alan Greenspan.

"Yeah, a guy who couldn't get a $16,000 loan wound up helping establish monetary policy for the United States of America," Brown says, with a chuckle.

And those banks that turned him down?

"Oh, they came around wanting to buy my company," he says. "I told them, 'Thanks, but no thanks.'"

PART III

Thunder at the Cape

8 JOANN MORGAN

Instrumentation Controller, Apollo Launch Control

JoAnn Morgan entered the firing room at Kennedy Space Center early one morning in 1964, took a seat at a console, and began to prepare for a test of the Saturn IB rocket, which eventually would launch the first manned Apollo flight four years later. She didn't even have time to plug in her headset before the launch director walked up to her and politely said, "We don't allow women in here."

Morgan, a twenty-three-year-old brunette, said there must be a mistake. She told the launch director she had been sent by Carl Sendler, an Austrian scientist and one of the leading members of NASA's rocket development team. The director shook his head the whole time Morgan was talking. "We don't allow women in here," he said again.

And he was correct. Females had never been permitted in the firing room, located about three miles from the launch pad. There wasn't even a women's restroom in the facility.

But times were about to change.

Morgan phoned Sendler and told him she was being asked to leave. Sendler responded: "Plug your headset in and get to work. Get that test done! I'll take care of that guy."

The director, new to NASA via the Navy, had no way of knowing Morgan's experience at Cape Canaveral, and apparently no one took the time to fill him in. She worked her first launch at the age of seventeen—five days after graduating from high school—as an engineer's aide with the Army Ballistic Missile Agency. She used a device similar to a telescope to help track the vehicle and assess whether the rocket's two stages separated properly. She was back at the Cape every summer during college, learning rocket science from Wernher von Braun's brilliant engineers. NASA hired her fulltime in 1962, the day after she graduated from Jacksonville (Alabama) State University with a math degree. Over the next two years, she proved to be one of Sendler's brightest, most dependable workers. That's why Sendler fought to get her a seat inside the firing room—he wanted the best people doing the toughest jobs.

"I found out later the conversations that morning in the firing room went all the way to the director of the space center, Dr. Kurt Debus,"

JOANN MORGAN

JoAnn Morgan, inside the Vehicle Assembly Building at the Kennedy Space Center, was a pioneer for females in the engineering world. [Courtesy NASA Kennedy Space Center, Florida]

Morgan says, nearly four decades later. "And it was made known to everybody there, from that point on, that if I was given a job and some men in the firing room didn't like it, they could just lump it."

Five years after that incident, in the summer of 1969, Morgan would be the only female inside Firing Room No. 1, located on the third floor of launch control, as Apollo 11 began its journey toward the first moon landing. She worked as an instrumentation controller on that historic flight, and on every moon mission thereafter.

"Five hundred men," Morgan says, "and me."

A naturally curious child, Morgan nearly blew up the family patio when she was in fourth grade.

"My parents gave me a chemistry set for Christmas that year," she says, "and I *loved* it. Chemistry was my dad's major in college, and when he was in the military he was in ordnance—so bombs and rockets and things like that were commonplace to him. And I was sort of the ringleader of the kids in my neighborhood [who were] experimenting with things."

One day she mixed some chemicals in a tin can and stuck it between two steps leading to the patio. "The last thing I added was what made it combustible, and I knew it," she says. "It wasn't supposed to go airborne, but bits and pieces did, and it wound up cracking the patio. Our housekeeper kept telling me, 'You're gonna be in *big* trouble.'

"But when my parents came home—and this was typical of my mother and father—they didn't fuss at me. My dad said, 'My goodness, how impressive! Look at that big ol' crack in our patio!' They never made me feel hesitant about trying things. Even when my sister and I tried cooking and it didn't go well, they would always reward us for trying and being adventurous and experimenting.

"They taught you to clean up your mess but you didn't get punished if there was collateral damage."

Morgan, born December 4, 1940, in Huntsville, Alabama, was the oldest of Donald and LaVerne Hardin's three children. She could read by the age of three and was already adding and subtracting when she entered first grade in Titusville, Florida. Her principal and teacher were so impressed, they promoted Morgan to second grade after just two weeks.

She was a math whiz throughout school. "My teacher caught me finishing my homework in class in tenth grade and started giving me extra work so I would have something to do at home," she says. "My senior year, I took trigonometry—there was no calculus class. Remember, this is 1958. But my teacher realized I wasn't being challenged, so he started giving me some calculus to help prepare me for college."

Morgan enrolled in the electrical engineering program at the University of Florida with a full academic scholarship, but she chose to stay only three years. "Their engineering program was five years," she says, "and I didn't want to be in school that long. I wanted to get out and go to work for NASA. Plus, my sister had transferred to Jacksonville State, and she wanted me to go with her. I stayed there long enough to get a math degree, and I was gone. I didn't even go to graduation. I headed straight to Cape Canaveral."

She took with her the free spirit her parents had encouraged, but also the manners of a typical Southern girl in the 1940s and 1950s. She said, "Yes, ma'am," and "No, ma'am," spoke only when spoken to in the presence of grownups, and obeyed the old creed: "If you don't have anything nice to say about someone, don't say anything."

"That's the one barrier of my upbringing that I had to overcome in the workplace," Morgan says. "During one of my first summers at the Cape, one of the [missile] tracking stations had been hit by lightning. I was supposed to look at every single thing that had been damaged and see if I could determine the path of the lightning and how much of the equipment was damaged. There was a telephone pole outside the equipment trailer with antennas and cables running up and down it. So I had to go up this pole and trace the lightning right into the equipment, make an inventory, [and] then go figure [out] how much it was going to cost to replace all the things that were burned or melted.

"I did all that and gave my boss, Jim White, the report. I went into a conference room with eighteen or twenty people, [who were] talking about the lightning hits and damage. One of them was a lightning expert who had come down from [the Marshall Spaceflight Center in] Huntsville. I never said a word during the meeting, and afterward [White] was furious with me. He said, 'You're the only person who saw it. Speak up!'

"If they had asked me, I would've told them. But nobody asked. I was still a teenager, waiting to speak when spoken to. I had to be taught by my boss to forget all that Southern training. He said, 'When you've seen something with your own eyes, it's your obligation to speak up. That's your job.'"

That lecture paid off when Morgan was assigned to Sendler's instrumentation team.

"Mr. Sendler never told me this, but other people told me," Morgan says. "He would say, 'JoAnn communicates the best. Men come in here and don't tell me the whole story.' I think he trusted me for that reason. I would tell him everything I thought he should be aware of, and that's one of the main reasons he wanted me on that console—I would keep him informed. If it was in the middle of the night, and the vehicle got hit by lightning and there was going to be a lot of rework, I would call him at home and tell him so he knew, when he was driving to work the next day, what to expect.

"Maybe I was a little chatty," she says, laughing, "but since he wanted to know, I left nothing out. I would even tell him if the test conductor was mad about something, or if he was upset with a particular person. I talked to him about 'people' things as well as the technical aspects. And Carl Sendler appreciated that because it put him in a position [of being] very knowledgeable when he talked to his boss."

As strong as Sendler's support was, he couldn't shelter Morgan from the inevitable problems of being the only woman among hundreds of men. And at that time in America, sexual harassment wasn't a familiar term.

Her first summer at the Cape, Morgan tried eating at the on-site cafeteria. "There were a lot of GIs in there, and there were a few 'woo hoos!' And that was fine," she says. "But when I got in line, several of them crowded around me and tried to get in line next to me. As we went through the line, some of the GIs would sort of bump themselves [up] against me. It was so uncomfortable, I didn't go back; I started bringing my lunch."

Even later, after she was married in 1965 and had become a regular in the firing room, Morgan would get obscene phone calls at her console during tests. And while trying to get her work done, she was being watched by men from Florida to Texas.

"The firing room always had an operational television," she says. "There were cameras in there so that people in other remote locations around the Cape and in Houston could actually see launch control and what was going on. I remember early on in the testing, before we'd flown Apollo 8, somebody would call and bring some report to the data room [which was] back behind the firing room. So I would get up, run and get it, and go back to my console. Finally, one of the TV technicians called me and said, 'Mrs. Morgan, I feel obliged to tell you this. When you get up and go out, there are some guys who call and ask me to zoom in on your fanny as you walk out. And they say your sure have a good looking rear end.'

"I thanked him and told him he needed to tell his boss. I said, 'What if the media got hold of this? Or worse, what if my dad or husband found out? They would beat the tar out of them.'

"So I started minimizing my time out of my chair. All these things were just a nuisance. They didn't enrage me. They just kept me from getting my work done, and I loved my work. I wanted to be 100 percent correct. I was so focused that when these little twiddly things would come up, it would be like a crab pinching you—just some little something to deal with. But I never let it deter me from my mission, which was working for NASA."

A launch pad is a lot like a storied football stadium—eerily silent and empty for months at a time. The thunderous roars of history and the nervous optimism about the next event are left to one's imagination until that day arrives when it all comes alive again, creating more drama and emotion.

That's the way it was at Kennedy Space Center's Pad 39A when Apollo astronauts sat atop a Saturn V rocket . . . waiting . . . waiting . . . and then riding the smoke and flames and thunder toward the moon. Casual observers would burst into tears; they couldn't figure out why.

"I think a lot of things bring out those emotions," Morgan says. "I think it's partly the fact that astronauts are onboard. It has something to do with that much rocket power being harnessed. But it's also connected to our sense of exploration and adventure, that sense of the unknown, [of] where these people were going.

"It's very American, but it's a universal thing. When people would come from overseas to watch a launch, they [would be] just totally blown away by it.

"I never got used to a launch. I never got used to the sheer night beauty of the Saturn V sitting on the launch pad, especially if they were tanking and you could see the misting of the gases. It was an absolutely gorgeous thing to see. Sometimes I'd get there a few minutes early before my shift. . . . I'd just stand in the parking area and look at it because it was so beautiful. It was amazing to think people could do this, could build this."

Pad 39A, completed in 1965, was the starting point for every moon mission except Apollo 10, which launched from Pad 39B because NASA wanted to make sure it could function as a backup. On a television screen, it looked simple enough: a hump of land, conveniently located beside the Atlantic Ocean in case the rocket went astray. But in many ways, the pad was as complicated as the spacecraft. The concrete structure, ninety feet above sea level, was filled with hidden sensors and circuits and cables, all critical to monitoring the space vehicle and the pad. Plus, it had a fifty-foot-deep flame trench that diverted the exhaust plume sideways. More than 50,000 gallons of water per minute dowsed the pad as soon as the rocket cleared the launch's umbilical tower, a unique facility of its own.

JoAnn Morgan worked her first unmanned launch at Cape Canaveral only a few days after graduating from high school. [Courtesy NASA Kennedy Space Center, Florida]

The tower, 402 feet high, was mobile and specially designed to withstand the heat and vibration of 7.5 million pounds of thrust. Ground crews attached the rocket to the tower in the Vehicle Assembly Building, Then, it was rolled out to the pad on a crawler transporter at one mile per hour. When in place, the tower was connected to the vehicle with heavy metallic umbilicals that carried cables for various instruments, lightning arrestors, and communication boxes.

This was Morgan's world. She understood the pad and the tower the way a medical specialist knows a particular part of the human body. "And I sort of looked at it like that," she says. "During and after a launch, we had to give the pad a good physical—take its vital signs and make sure things were the way they should be.

"Remember, we were doing three and four launches a year, at one point. We didn't need the pad to burn up. So it was important to put the fires out, know where the damage was, give post-launch reports, and let people know the condition of everything that supported a launch.

"For instance, we used the same cable runs and cross-country lines from the big tanks that stored liquid hydrogen and oxygen for every launch. They're down at sub-zero level because the cryogenic fluids [they carry] are minus-300 degrees. If something hot flopped off onto one of those cold lines, it could break very easily. All of those things were of big concern. We didn't have time for the pad to go through six months of refurbishments."

About two weeks before the Apollo 11 launch, Morgan learned she had been chosen to work the prime shift, which meant she would be at a console in the firing room, with dozens of people at various sites around the Cape reporting to her, as Neil Armstrong, Buzz Aldrin, and Michael Collins departed for the moon.

She was excited for two reasons: One, Sendler having made her a part of the A team for a launch of such historical significance said something about her ability. Two, she didn't have to go through the tedious hours of loading the rocket with liquid hydrogen and oxygen. She had pulled that early shift for Apollo 10, and it was stressful work.

"That's the hardest part, getting the propellants loaded and stabilized," she says. "You're looking for leaks or fires. You're watching the weather because you don't even flow those propellants if lightning is within five miles of the pad. So that is the dangerous time for the pad itself.

"But for my Apollo 11 shift, I would go on duty three hours before liftoff, just about the time the astronauts were being put inside the spacecraft. We had to make sure the communication system was working perfectly, and we really worked with the people at Houston and Goddard [Space Flight Center in Greenbelt, Maryland] on that, plus the tracking systems for the vehicle. We were checking the guidance and control systems."

She was confident. "I knew I was going to do my job well because I'd worked there off and on for eleven years and full time for seven."

For the most important day of her career, Morgan struggled with just one thing: What to wear to the launch.

"Honestly, by 1969, the only time I thought about being the only woman in launch control was when I got dressed for work," she says. "For the Apollo 11 launch I thought, 'You know, all those men are going to be wearing their coats and ties. What can I wear that will blend in?'

I certainly didn't want to go in there with a bright yellow dress and [bring] attention to myself.

"So I chose a navy Lacoste dress, with a thin red strip in it that you couldn't even see from ten feet away. And since I was a Florida Gator fan, I was happy to have the little alligator figure on it."

The day Apollo 11 went up, launch control was filled with NASA brass, launch specialists, and contractors—people like von Braun, head of Marshall Spaceflight Center in Huntsville; Kurt Debus, director of Kennedy Space Center; Deke Slayton, head of flight crew operations at the Manned Spaceflight Center in Houston; and Alan Shepard, America's first man in space and head of the astronaut office in Houston.

The countdown, which commenced at T minus twenty-eight hours, went smoothly, and at 9:32 a.m., the Saturn V roared to life and slowly made its climb past the tower. "Those thirteen seconds or so that it took for a Saturn to get above the tower seemed like thirteen minutes," Morgan says. "It seemed like it was moving so slowly."

As the vehicle gained speed and performed its automated roll and pitch programs to reach the correct trajectory, almost everyone in Firing Room No. 1 rose from their seats to watch through the large window in the back of the room. Launch specialists, contractors, and NASA brass stood and stared. But not JoAnn Morgan. She remained seated and focused. There is no disputing this; NASA has a picture to prove it.

"The reason I had to stay seated was my instrumentation team was scattered throughout the entire center," she says. "I had people at observer sites with binoculars [who were] looking at the color of the rocket's exhaust to determine if the burn was appropriate. We had people scattered all over the area feeding reports like that. And people were reporting to me if there were fires on the pad, if there were fires in the areas where maybe some debris had landed. I was receiving reports about possible equipment burning [or] cables and sensors burning. I was having to process a lot of information in a short amount of time.

"I wanted to watch the vehicle flying to the moon, but I was too busy for more than just a glance."

She was one of the busiest people at launch control over the next two hours. "All the other people at the consoles started emptying out about thirty minutes after the launch. I had to get my damage reports in, how the systems performed on the ground, how much time it would take to get ready for the next launch. All of my people were looking for the something [that was] not normal. Maybe we had scorched some cables in one area that are going to have to be replaced. Or maybe we lost two communication boxes. Or the lightning antenna got blown away."

Apollo launch control as Apollo 11 rises from Pad 39A. Notice JoAnn Morgan in the center of the picture. She is the only female and one of the few who stayed in his/her chair. [Courtesy NASA Kennedy Space Center, Florida]

The Apollo 11 crew had already made the Translunar Injection burn by the time Morgan finished her reports. She walked out that day with Slayton and Shepard, who were about to board a jet for the Manned Spaceflight Center in Houston. Morgan and her husband, Larry, a high school math and science teacher and bandmaster, were headed for a much-needed vacation along the Florida Gulf Coast.

She had worked tirelessly since taking the job with NASA. In 1967, while trying to help get the Apollo program off the ground, Morgan collapsed in an elevator at the Kennedy Space Center. She was pregnant and began hemorrhaging. And as she prefers to say now, that's the day she and Larry became "parents of an angel." She miscarried.

"I'm sure it was partly due to stress," she says. "It was one of those pregnancies where I was working right up to the day I lost the baby. They got me to the hospital, but it was too late."

During the summer of 1969, she had worked twenty-eight straight days prior to the launch of Apollo 11. So as Neil Armstrong, Michael

Collins, and Buzz Aldrin headed for the moon, she and Larry relaxed at Florida's Captiva Island.

"It took a couple of days of running around on the beach and going out in the boat for the launch to really sink in," she says. "I had been *so* focused. But it suddenly hit me just how historic the launch was. . . . Being the only woman there, how historic it was for me and women and engineering. We could always say at least one woman was part of that launch team. I look back at it now and say without reservation that it was the greatest launch team in the world. It really was. Nobody has ever touched it."

In their motel room, Larry and JoAnn watched on television as Armstrong and Aldrin made the first human steps on the moon. "We had a bottle of champagne," she says, "and I remember that it was in the wee hours of the morning and how unusual it was for a TV station to even be on that late at night—back then most of them went off at midnight or shortly after.

"And then the whole thing hit me again as I watched Neil and Buzz walk around. I was like, *Ahhhh!* I helped get them up there!"

Morgan spent forty-five years with NASA before retiring in November 2003. Following the moon landings, she worked the Skylab and Apollo-Soyuz missions and helped develop the launch procedures for the shuttle program. She was the first female senior executive at Kennedy Space Center, earning four Exceptional Service Medals, and was inducted into the Florida Women's Hall of Fame.

She and Larry divide their time these days between a beach house in Florida and a ranch in Montana.

With the same passion she used to approach her job, Morgan speaks out about the current status of America's space program: "We're way behind where we should be. Way behind what our potential is. I look up at the moon today in amazement that we were able to do what we did with such little technology, compared to what is available today. But we're just not using it. We've totally lost our expectations of young people. We quit toughening kids up for life and started babying them at some point. Playing video games all the time doesn't toughen a person up very much.

"I think it's one of the saddest things I've ever seen—next to the deaths of the astronauts and some of the ground crew who were killed in accidents. Historians will look at Apollo and wonder why we didn't keep going to the moon and beyond, and they'll look at political and financial and cultural issues. They'll try to deduce the factors that contributed to it, and there was certainly more than one factor. And they will assess it a little more purely than people [did] while [it was happening]."

She is frustrated with the lack of women interested in engineering. "Females make up only about thirteen percent of the engineering work force," she says. "Engineering is a tough field for women to get into and stay in because of the demanding nature of it. I won't say men tend to be more arrogant than women, but there's a confidence level that goes with being a scientist and engineer that is required, and it's not a quality that's inherent in women's behavior. These TV shows have all these pushy women lawyers and doctors. They may have 'em out there in the legal and medical professions, but you just don't see them in the engineering field."

Morgan says she sometimes believes her time spent helping Americans get to the moon was a matter of "just sheer luck." She explains: "I was lucky to be the oldest child of parents—and grandparents—who encouraged me to be fearless. Lucky to have a mentor like Carl Sendler and to work in an environment that would challenge me. And then lucky to be at Kennedy Space Center and continue to be rewarded with newer challenges through jobs and promotions.

"Hey, I was in the right place at the right time."

Morgan means she was lucky overall and particularly during one incident. On January 27, 1967, Morgan arrived at 6:30 a.m. to find out that the Apollo 1 crew—Gus Grissom, Ed White, and Roger Chaffee—would be at the Cape that afternoon for what was known as a plugs-out test.

"It's where you pop the umbilicals away from the vehicle, and the astronauts wanted to know what it felt like," Morgan says. "When [the umbilicals are popped], the vehicle sort of sways a little, and they wanted to be used to it for the launch. And there were other things they wanted to see too. When power is transferred from the ground to the onboard systems in the cockpit, it's sort of like disconnecting the charger to your cell phone—you want to see if it stays lit and works properly. They were that way about their spacecraft. They wanted to know what it was going to be like."

Morgan knew all three astronauts. "It was a small enough work force, at the time, and not that many people [were] out in the blockhouse, and the astronauts would always wander in and talk to people. I'd known Gus for a long time. I worked his Mercury flight when I was still in college."

The test had one bug after another. "I was calling my boss [Jim Koontz] and keeping him updated. When 4:30 rolled around, and we still hadn't gotten to T minus zero, [Koontz] relieved me and made me go home."

Two hours later, the three astronauts perished in a fire inside the command module. Koontz phoned Morgan at home and told her.

"He didn't want me to hear it on the news or get up the next morning and read it in the paper," she says. "He wanted me to come in the next morning. [We] had to impound data. We had a procedure to follow; all the documents of the test had to be packaged up.

"I look back on it now, and I know if my boss hadn't relieved me, it might have changed my career. He certainly was never the same. He never wanted to go and sit on the console during those types [of] tests anymore. It devastated him to be the person who was on duty when it happened, and a lot of people who were in the blockhouse that night were affected. They were on the communications loop. They heard what went on. They were there when the bodies were pulled out.

"I just can't help but believe if I had been there, it might have taken the steam out of me. The fact that I didn't have to *live* that [event] probably saved my career. Again, I was just lucky."

9 JOE SCHMITT

Suit Technician

Joe Schmitt was getting a haircut in Pasadena, Texas, during the summer of 1999 when he glanced up and saw a television news report. Astronaut Gus Grissom's *Liberty Bell 7* Mercury spacecraft had just been recovered after thirty-eight years on the floor of the Atlantic Ocean.

"Hmmph," Schmitt said to no one in particular. "I put Gus in that capsule."

The shop fell silent, except for the steady voice of the TV anchor. A man sitting across the room lowered his newspaper a few inches so that his eyes could study the person who had made such a bizarre claim. A customer one chair over stared hard at him. And Schmitt no longer heard scissors snipping away at his gray hair; the barber was stunned too.

Schmitt was never one to volunteer information about his twenty-four years with NASA as a suit technician. He didn't talk about helping men dress before they left for the moon, or being the last to shake their hands before liftoff, or that his face was on two Norman Rockwell paintings. He figured it might come across as bragging, and he never cared much for arrogant people.

His words that day at the barbershop just slipped out. He was thrilled they had rescued Gus' capsule, which sank when a hatch accidentally blew open following his fifteen-minute suborbital flight in July 1961. He knew how happy Grissom—who died in the Apollo 1 pad fire in 1967—would have been to see his old ship back home and on its way to being restored, the object of thousands of admiring eyes. And so he had spoken without thinking.

"I didn't say anything else, and everybody finally went on about their business," Schmitt recalls now, with a chuckle. "I'm not sure what they thought. They didn't know anything about me, or at least what I used to do. And there's no reason they should have. I always [say] I'm not famous, but I sure was around a lot of people who were."

Schmitt's extended family came from Germany near the end of the nineteen century and settled in O'Fallon, Illinois, a small town in the southern part of the state. They were hard workers, each specializing in a particular trade. One of his uncles was an expert trapper, another a shoemaker, another a butcher. They raised pigs and ate nearly every part of them. Times were tough.

And they got a lot tougher. Schmitt's father, Benjamin, was a policeman in O'Fallon. The father of seven was pulling duty at the town's train depot

JOE SCHMITT

Joe and Libby Schmitt live a quiet life these days at their home outside Houston. Schmitt was a suit technician during Mercury, Gemini, and Apollo. [Courtesy NASA]

in April 1916 when a World War I veteran drifted into town, possessing a pistol and a twisted mind. He shot the first person he saw when he got off the train—Benjamin Schmitt.

Joe Schmitt was just ten weeks old when his daddy was murdered. His mother, Addie, never remarried and raised her children on the money she earned by washing other people's clothes.

"I had this little four-wheel wagon, and I can remember delivering the clothes back to the people she had washed for," Schmitt says. "All of us kids did whatever we could to make it easier on our mother."

Afternoon and weekends throughout his school years, Schmitt cleaned spittoons and shined shoes at his brother-in-law's barbershop. He earned a dime a shine, and he saved for three years until he could finally buy his mother a stove.

"Ours had broken, but we got by," Schmitt says. "I think the new one only cost about $30, but that was a lot of money back then."

College was financially out of the question for Schmitt. "I couldn't get a job—nobody could because of the Depression," he says. "So the high school said it was okay if we came back and took some extra courses. I took typing and some kind of business course—which I never used. But while I was there, the principal noticed that I was mechanically inclined. He had heard about some of [the] home projects I'd done." (Word travels fast when someone in their teens builds a motorized Ferris wheel out of Tinker Toys.)

"The principal asked me how I would like to go into the [Army] Air Corps," Schmitt says. "I told him I'd like it fine. So he wrote a letter to the commanding officer at Chanute Field, explaining my situation. Lo and behold, they accepted me."

In 1936, he began training as an airplane mechanic, working on bombers such as the B-6 and B-39. "But things were kind of slow during peace time," Schmitt says. "So they let me take an [aircraft] instruments specialists course. And I also had time to take a parachute riggers course, and part of that was learning about aircraft clothing repair. Looking back now, I guess taking that course is what allowed me to get into the space program."

He left the Air Corps in 1939 and joined the National Advisory Committee for Aeronautics (NACA) and worked as an instruments specialist based at Langley Research Center in Virginia. Schmitt helped install the flight measurement instruments onboard the XS-1 rocket plane, in which test pilot Chuck Yeager first broke the sound barrier on October 14, 1947.

It was the first of many firsts in which Schmitt would be involved.

When the NACA became NASA in 1958, he was asked to take a job in the crew systems division. "They saw I had done a little work in aircraft clothing repair and asked if I wanted to work on spacesuits. I said sure."

Schmitt remained at Langley to study the makeup of the Mark IV pressure suit that was being used for high-altitude Navy test flights.

At that point, scientists weren't exactly sure what kind of suit was needed for a human to survive in space. But this much they did know: At sea level, the body is squeezed in all directions at a pressure of 14.7 pounds per square inch. This allows blood to flow and provide fuel and oxygen head-to-toe. But pressure decreases at higher altitudes, which decreases blood flow. Plus, the higher one goes, the less oxygen is available. In space, there is no oxygen, and temperatures range from plus or minus 200 degrees. The body's organs, which rely on moving parts—the heart, for instance—will shut down.

B. F. Goodrich was awarded the contract in July 1959 to make the spacesuits. Schmitt traveled to the Goodrich plant in Akron, Ohio, and monitored the design and manufacturing. "I helped do the suit fitting stuff there and learned how to check out a particular suit," he says. "I always liked to learn new things, and I enjoyed a challenge. This was right up my alley."

In September 1960, Goodrich delivered twenty-one flight-ready suits to Cape Canaveral—three each, tailor-made, for the Original 7 astronauts. Included in this were one training suit, one flight suit, and a backup. They shined like aluminum foil and became pressurized only in case of an emergency—the spacecraft cabin would be pressurized, instead, to five pounds per square inch. Collectively, they cost approximately $100,000, or about the same price as just one of the suits used to walk on the moon. The Mercury suits consisted of an inner layer made of neoprene-coated nylon and an outer layer made of aluminized nylon. The astronauts found them uncomfortable, but efficient.

To better understand his job, Schmitt donned a spacesuit for a chamber test at Langley. "We went up to about 30,000 feet, and that's where you learn just how immobile you feel when you're pressurized," he recalls. "You just feel like you're inside a man-shaped balloon. It's not a good feeling. And an astronaut had to be able to reach switches and gauges without any problems. So getting the suits right for each astronaut took some time."

As the chief technician, Schmitt maintained each suit, checking the stitching, performing pressure checks, and making any fitting adjustments necessary. The suits were kept locked away in a special dressing room at Cape Canaveral.

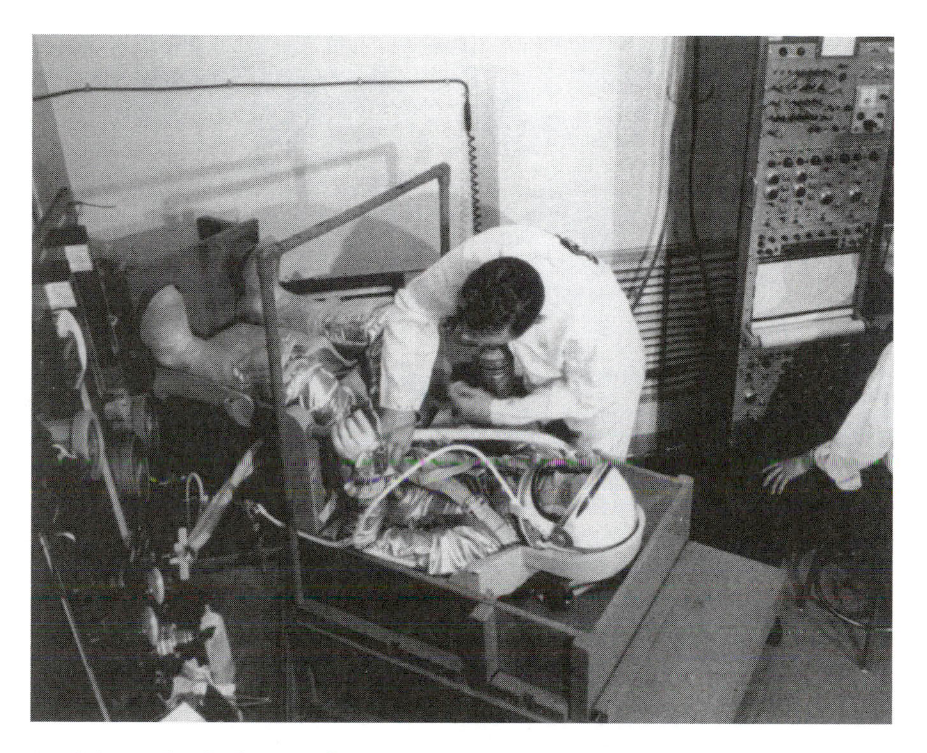

Joe Schmitt checks the suit of Mercury astronaut Scott Carpenter during a training session. [Courtesy NASA]

"I rigged up this console where I could check the suit for any problems," Schmitt says. "I was quite proud of it. I mounted various flow meters and pressure gauges. It was pretty primitive looking back now, but it served its purpose at the time."

He laughs. "I was doing a pressure check on Wally Schirra's suit one day. Pushed it up to five pounds per square inch—sort of a max-out test. And the thing blew. Sounded like a shotgun went off. We wrapped it up and sent it back to the factory."

Schmitt didn't need a gauge to measure the pressure involved in his job. He knew that one hole the size of a pinhead could cause instant death in space. Schmitt didn't get a minute of sleep the night before Alan Shepard was to make America's first manned venture into space, a fifteen-minute suborbital flight in May 1961.

"I was staying at a hotel near the Cape," he says, "and to be honest I was scared to death that I was going to forget something in the procedure. So I took out one of those laundry forms they have in motel rooms, and on the back I wrote out the checkout procedure step-by-step. I kept it with me, and I told [pad leader] Guenter Wendt that before he closed

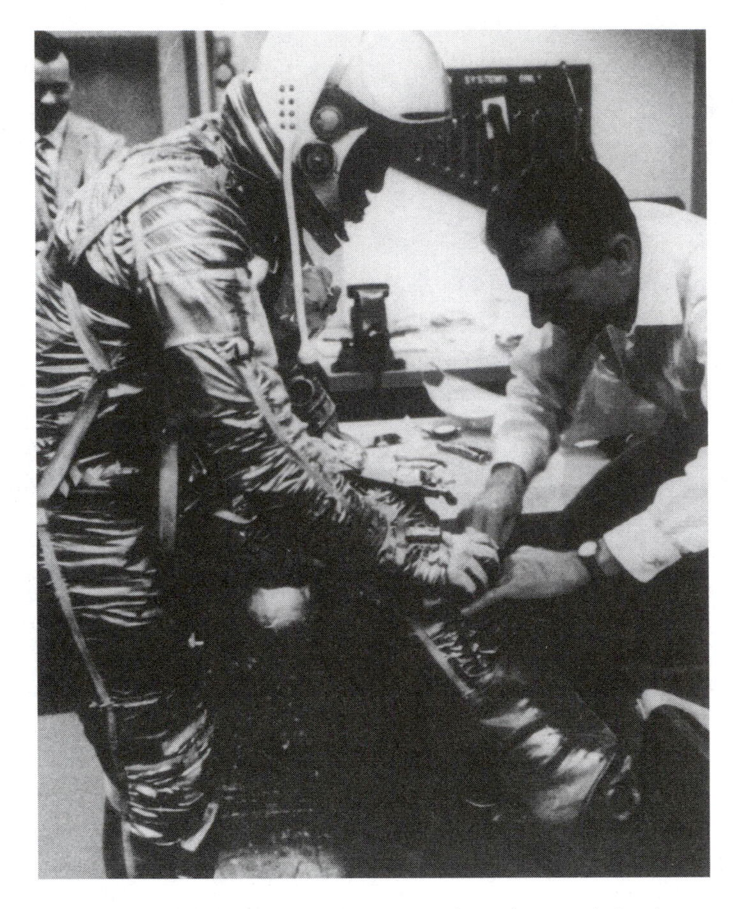

Joe Schmitt helps Mercury astronaut John Glenn with his boot. [Courtesy NASA]

the hatch, I wanted to go right down my checklist list again. And I did that on every flight after that."

Over the next five years, Schmitt worked five of the six Mercury missions and six of the ten Gemini flights. He saw a lot of anxious eyes as he strapped numerous friends into spaceships that were going a long way from home. And he did it with one thing in mind: Help America meet the challenge set by President Kennedy to land a man on the moon during the 1960s. "All of us had tunnel vision," he says.

As Apollo 11 prepared to launch in July 1969, just five months from the deadline, Schmitt could sense the urgency felt throughout NASA to get this mission done. "So many people had spent so many hours doing the very best they could to get that mission off on time, and I don't think people understand just how much pressure we were all under," he says. "There were people in the program who had more important jobs than mine. But every person took [his or her] individual duties seriously. Nobody wanted to be the one who screwed up. I sure didn't."

Schmitt reported for duty on the third floor of the Manned Space Operations building at 3 a.m., more than six hours before Neil Armstrong, Buzz Aldrin, and Michael Collins were scheduled to blast off on Apollo 11. He and four other technicians ran the suits through routine oxygen and communication tests. They checked the zippers on the backside of the suits to make sure they were intact and working smoothly. They loaded each astronaut's pockets with flashlights, handkerchiefs, pencils, and scissors strong enough to cut wire.

Armstrong, Aldrin, and Collins arrived around 5:30 a.m. after the standard astronaut breakfast of steak and eggs. Schmitt knew all three extremely well and was familiar with their personalities when there were no cameras or reporters or bosses around. He studied them closely that morning to see if they might be nervous or hesitant about the mission. All he could sense was a quiet confidence.

"Neil was always very passive about things," Schmitt says. "Even when I'd suit him up before tests, it was like his mind was way ahead of what we were doing. It was like he was already thinking about what he was going to be doing inside the spacecraft. If he had a problem, he'd let you know it. But he was pretty good about just going along with things.

"Buzz . . . now he wanted everything just right. I remember on Gemini 12, he had to go outside the spacecraft and back to the service module on a little EVA. Back there, we had mounted this board that had various valves and switches—things they would have to manipulate in the Apollo program. See, we were always preparing for Apollo.

"Anyhow, prior to the flight, we gave him a choice of gloves in his size range, but he could never get one that felt good on his right hand. He wound up using the right glove from another astronaut's training suit. But since it has been used in training, it wasn't qualified for flight. So we had to redo the whole glove, get it qualified, and bring it up from training status to flight status. That's what he wanted, so that's what we gave him. And that was our job, to make them as comfortable as possible so they could do their very best.

"Mike Collins was just a real pleasant guy. I really enjoyed his company."

During suit-up, Schmitt and two other technicians made sure everything fit properly.

"They put on long underwear, a jock strap and a urine collection device," Schmitt explains. "Then the doctor came in and attached the bio-medical sensors to their bodies. They would check the sensors, make sure they were working properly, and then we'd help them get into their suits. But before we closed the big zipper on the back, the doctors did one more check of the bio-meds, the reason being those suits were fairly heavy, and it was easy to disturb some of those sensors."

The Apollo suit, made by the International Latex Corporation, was light years ahead of the old Mercury get-up. It was still pressurized to five pounds per square inch and tailor-made to fit each astronaut, but it was water cooled and made of nylon coated with Kapton for heat protection and Teflon to prevent tears and scrapes.

"That suit was a real piece of work," Schmitt says. "I don't know if it was a lot more comfortable than the Mercury suit, but it certainly offered more mobility. The astronauts could reach around much better, and that was really important in Apollo because they had so much more maneuvering to do."

About three hours before launch, the suits were hooked to portable ventilators that fed the astronauts pure oxygen. "You wanted to get as much nitrogen out of their systems as possible before liftoff. If you didn't, they could get something equivalent to the bends—like a diver who comes up too quickly," Schmitt explains. "So the big deal for us, after getting them suited up, was to maintain this pre-breathing routine. If it was interrupted for some reason, you'd have to scrub the mission."

As the astronauts exited the MSO building and headed for the transfer van, which would make the twenty-minute ride to Pad 39A, Schmitt walked a few paces behind, dressed in a white "ice cream" suit. "I've never seen that many photographers in my entire life," he says. "It was their last chance to get a picture of these 'moon men' and they certainly didn't waste the opportunity."

It was a two-elevator ride up to the command module, 323 feet high. The first elevator took Schmitt and the astronauts forty feet to the base of the Saturn V. The second elevator completed the climb to the white room—the ultra-clean work area surrounding the spacecraft. Only Schmitt, Wendt, another suit tech, an inspector, and backup lunar module pilot Fred Haise—who had been pre-setting nearly 700 switches inside the command module—were allowed in the white room at this point in the countdown.

Procedures had changed since Mercury and Gemini. There used to be endless chatter between the astronauts and the pad crew, a few jokes here and there to lighten the mood. But the Apollo 1 fire during a test

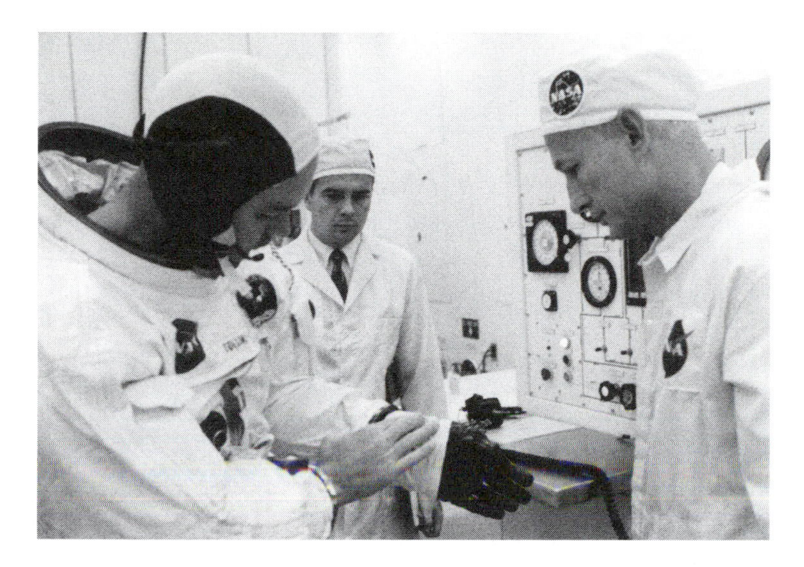

Joe Schmitt assists Michael Collins, the Apollo 11 command module pilot, on the morning of the launch. [Courtesy NASA]

when Grissom repeatedly complained of not being able to hear the people inside the launch control center ended that. Poor communication had nothing to do with the death of Grissom, White, and Chaffee, but NASA officials wanted chitchat kept to a minimum.

"What they did was take the microphones off our headsets," Schmitt says, laughing. "We could hear everything being said between the spacecraft and the launch control center; we just couldn't add anything to the conversation. And I thought it was a good idea because the astronauts needed to be able to talk with the flight director and not be interrupted if there was a problem with systems or the weather. So I had to work out a set of hand signals with the astronauts. They would give me the 'Flying-O' signal if everything was okay, or they would point to whatever they needed help with."

Still, the astronauts couldn't pass up the chance for a good laugh or two. That was their nature—greet danger with laughter.

Armstrong presented Wendt a small piece of paper that read "Space Taxi Ticket, good between any two planets." Collins handed Wendt, who was an avid outdoorsman, a brown paper bag containing a six-inch fish nailed to a board with the inscription "Guenter Wendt, Trophy Trout." Aldrin was a bit more serious. He gave the pad leader a Bible inscribed: "On permanent loan to Guenter Wendt."

Wendt shook their hands and moved aside. It was time for Schmitt to do his thing.

Armstrong entered the spacecraft first and slid into the commander's couch, on the left side.

"I [followed] him in and [crouched] just inside the spacecraft," Schmitt explains. "Guenter was standing just outside the hatch making sure my back didn't hit any of the switches on the instrument panel. First thing I did was connect Neil's communications [hose] to a spot on the front of his suit. I wanted him to be able to talk with launch control. In addition to that, I had to connect the plug for the bio-meds so that the doctors could monitor their vital signs.

"The portable ventilators were still pumping oxygen, and they had long hoses so that Guenter could stand just outside the hatch and hold the ventilator until I could get everything hooked up in the spacecraft. I connected the oxygen hose to Neil's suit and turned off the ventilator.

"Then I started buckling the harnesses—just like the seatbelts in your car, only a lot more sturdy and a double set of straps. When Neil felt good and everything was connected properly, he gave me the 'Flying O,' and I stepped out of the way so the next astronaut could enter."

On any other Apollo mission, the lunar module pilot would have been next to enter the spacecraft and slide into the couch on the right. But Aldrin, the lunar module pilot on Apollo 11, had trained on the Apollo 8 backup crew as the command module pilot (along with Armstrong and Haise). He was already familiar with the launch duties of the center couch. So when training began for Apollo 11, mission planners decided to keep Aldrin in the middle and train Collins for the right seat.

In a matter of minutes, all three astronauts were in their couches, snug and ready for the ride of their lives. Schmitt looked them square in the eyes, as he shook their hands and offered the only four words that seemed appropriate: "Have a good flight." Each responded with a Flying O.

"That's when the closeout crew entered the spacecraft area, and we had to collect all the loose gear—the hoses, the rubber slippers the astronauts had worn over their boots on the way to the pad."

For a precious few moments, Schmitt took the opportunity to look *down* at the monstrous Saturn V, the most powerful rocket ever built. And he couldn't believe what he saw: In the middle of a Florida summer, it was snowing. "Little flakes of ice were falling off the cryogenic tanks," Schmitt says. "That was *very* cool."

Schmitt and the rest of the white-room workers spent the final minutes of the countdown a couple of miles from the pad in what was known around the Cape as a fallback area—probably the closest view anyone had of the launch. "We were kept there because, if the mission was scrubbed for any reason, we would [be able to] get back into the van and hurry over to get the astronauts out of the spacecraft. We didn't want them just sitting up there for no reason."

The astronauts didn't sit long on this particular morning. At 9:32 a.m. Florida time, Apollo 11 lifted off. Four days later, in an area of the moon known as the Sea of Tranquility, Armstrong and Aldrin touched down.

If Schmitt could've seen them at that moment, he would've given them a Flying O.

"Everything just fell in place for us," he says. "That's all we can say. It's mind-boggling when you step back and look at what we did in such a short period of time. It wasn't but eight years earlier that I strapped Alan Shepard into a capsule and [hoped] he could survive fifteen minutes in space. I'm proud I had just a little part in it all."

Schmitt wasn't finished. He worked one more moon mission: Apollo 15, another landmark flight, was the first to utilize a car on the moon, and it allowed Dave Scott and Jim Irwin to do some serious exploring.

He suited up Pete Conrad for the Skylab mission, John Young for the first shuttle flight. Then in 1983, which was his twenty-fifth year with NASA, Schmitt retired. It wasn't forced. Nobody "strongly suggested" it. He just walked away to spend more time with his wife Libby and their children and grandchildren. He gardens. He does woodwork. Says Schmitt, "I guess I got my job with NASA because I was good with my hands, and I still like using them."

Astronauts, for the most part, are not sentimental people. They do their jobs and quickly move on to something else—another mission, another career. But the soft-spoken, kind-hearted Schmitt—the last person most of them saw in, perhaps, the most intense moments of their lives—had a way of chipping through their crusty personalities.

When John Glenn returned from his three-orbit Mercury mission in 1962, he gave Schmitt a gold medal with his initials imprinted on the back. "I found out later he took about ten of those on the flight with him," Schmitt says. "He gave one to President Kennedy, six to the other original astronauts, a couple to the NASA higher-ups, and one to me. I couldn't believe it. And I've still got it."

Collins, who also flew on Gemini 10, called Schmitt aside one day and handed him a small piece of the spacecraft's heat shield. Such items were few and precious. But he told Schmitt: "I can't think of anyone in NASA who deserves this more than you. Thanks for all you do for us."

Schmitt still has that memento too, along with a U.S. flag that flew aboard Apollo 11.

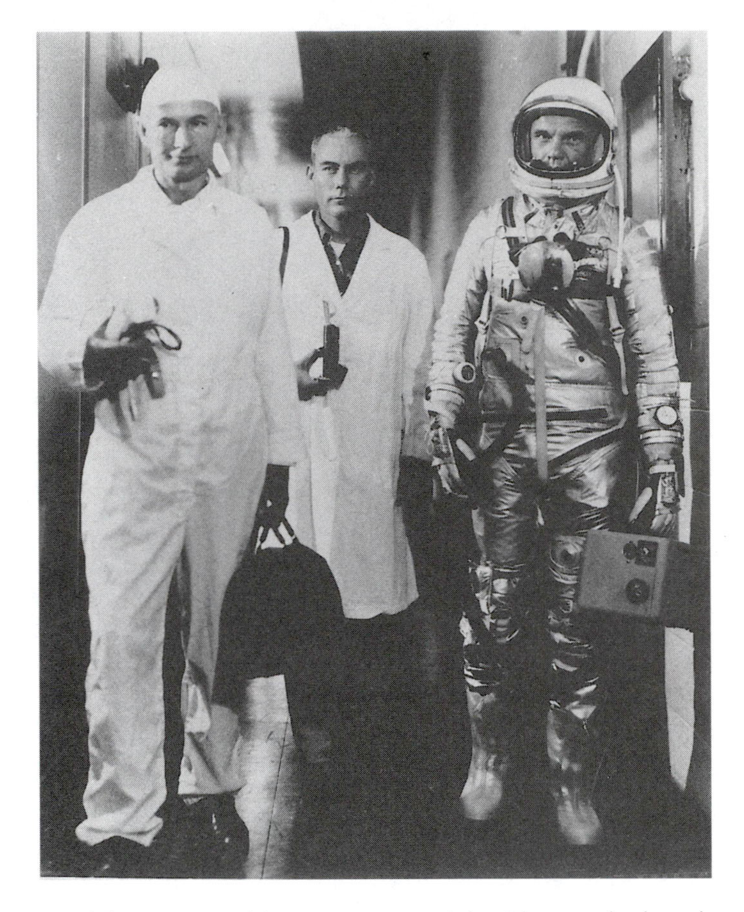

Joe Schmitt escorts Mercury astronaut John Glenn to the launch pad. Glenn was the first American to orbit the earth. [Courtesy NASA]

"It's always nice to be appreciated by the people you work with, no matter what you do," he says.

But one thing always bugged him: Why, Schmitt wondered, did Norman Rockwell feel compelled to include him in the paintings he did of Gemini 3 and the Apollo program. In both, the astronauts are in the forefront, and Schmitt is standing behind them.

He finally got a chance to ask the legendary artist. Rockwell answered simply, "Because you were always there."

10 JACK KING

Voice of Apollo/Saturn Launch Control

On the morning Alan Shepard was launched into space, Jack King saw the streaming fire beneath the rocket, heard the rumbling of powerful engines, considered the fact there was a person on top of that thing, and realized the historical impact of an American *finally* going up. It brought tears to the eyes of NASA's chief public information officer.

"When you're close to something like that, and you know what everyone has gone through to make it happen, it affects you," King says.

Eight years later, in 1969, as the voice of Apollo/Saturn Launch Control, King counted down the liftoff of Apollo 11—America's first stab at landing a man on the moon. His words, delivered with a brash Boston accent, were heard on television and radio throughout the world.

By his estimation, King had done hundreds of launches by then, including all the manned flights since Gemini 4. They never became routine, whether it was a satellite or a human onboard. But since the Shepard flight, and especially after becoming part of the launch team, King had made it a point to keep his emotions intact. Provide information quickly and succinctly, he always reminded himself, and don't become part of the story.

Yet as the final anxious seconds of the Apollo 11 countdown burst into a thunderous roar that rattled furniture for miles down the coast of Florida, Jack King's emotions nearly got the best of him.

"*Six, five, four, three, two,*" his voice began to crack, "*. . . one, zero. All engines running.*" At this point, there was a slight pause as the Saturn V, carrying three astronauts, slowly rose from the launch pad with force

JACK KING

Jack King, voice of Apollo Launch Control, watches as Apollo 11 leaves on its way to the moon. [Courtesy Jack King]

and grace once thought to be impossible partners. Somehow, King gathered himself. "*Liftoff,*" he said firmly, clearly. "*We have a liftoff, thirty-two minutes past the hour. Liftoff on Apollo 11.*"

More than three decades later, King isn't bothered by the emotional slip. "Normally, I have about five voices in my ear during a launch," King says. "People like the launch director, the launch operations manager, the Saturn rocket test supervisor. I'm listening to them, and then feeding the information to the public. But for that flight there was an additional voice—mine. And it was saying, 'My God, we're going to the moon.'"

At a thirtieth anniversary celebration of the Apollo 11 lunar landing, someone kidded King about his quaking voice. Charlie Duke, who explored the moon on Apollo 16 and was the capsule communicator as Neil Armstrong and Buzz Aldrin made the historic first landing, quickly came to his defense. Duke said to King: "I understand you getting emotional. Heck, I couldn't even say 'Tranquility Base' when Neil and Buzz landed. It came out 'Twanqwility.' That flight got to all of us."

King arrived at Cape Canaveral in 1958 to open an Associated Press (AP) news bureau. On the heels of Russia's successful launch of the unmanned satellite Sputnik in 1957, NASA was formed to get Americans into outer space as quickly as possible. "The AP decided we needed to start paying more attention to what was going on down at the Cape," he says, "and my bureau chief nominated me."

King, the son of a Boston sportswriter, was twenty-seven and news hungry. He had covered a little bit of everything in his hometown—shootings, car wrecks, and Red Sox baseball. "Saw Ted Williams hit his last home run," he says proudly.

He wasn't sure what to expect when he arrived at the Cape, a point on Florida's east coast that juts out into the Atlantic Ocean. This is what he found: Land with a lot of scrub brush and hardly any people. It had a desolate feel—perhaps the perfect place to begin a journey to the moon.

The Cape had been selected in 1949 as the U.S. military's first long-range missile test site, but the area was known as Cape Canaveral long before the government moved in; maps from as early as 1564 included Cape Canaveral, and some referred to it as "Cape of Currents." It was an ideal spot to launch missiles and rockets. They could begin their journeys out over the Atlantic, away from populated areas. It was accessible by land and sea. Plus, the Cape was located near the equator, which meant rockets would require less thrust, because they could take advantage of the earth's rotation, which is greater at the equator.

The first launch at Cape Canaveral occurred on July 24, 1950, when a modified German V-2 rocket soared about ten miles up before falling into the Atlantic, a mild success. By the time King arrived in 1958, the Air Force was firing off a steady stream of test missiles.

"They were top secret," he says. "If they told us anything at all, they would announce [to the media] that there would be a launch on, say, Thursday. And we would have to be there an hour early to take a bus out for viewing. We weren't to write anything about it until one minute after liftoff."

But the Air Force didn't publicize all launches. "As good newsmen, it was our responsibility to try and find out about them," King says. "It was a very small community at that time, so we knew where all the test conductors lived. If [a] car wasn't in the driveway at 7 p.m., that could mean—combined with a lot of other information we had gathered—that they were out there preparing for a secret launch. We caught them most of the time. They didn't get away with much, and I was very proud of that.

"There was one missile called the Bull Goose," King says, referring to a non-lethal cruise missile intended to send false electronic data to enemy defense systems and divert attention from manned bombers. "And my office in 1958 was at the Van Guard Motel. I came in one night, and the desk clerk told me, 'The Goose man just checked in.'"

King had no idea what a Goose missile was, but he researched it and realized it was a story. He phoned an AP photographer in Jacksonville and told him to hurry down and get pictures of it.

"In those days, they would raise a big red ball and flashing lights would occur a few minutes before a launch," King recalls. "The Goose didn't have the high-arc launch that you usually see. They could kind of sneak it off. Anyway, they launched it, and I wrote about it.

"General Don Yates was the commander of the Air Force Missile Command Center, and when I first came down, I told the public relations guys I'd like to meet him. They said, 'Sure, we'll take care of it.' But they never did. Soon as the papers hit the streets with the Bull Goose story, his secretary was on the phone and said, 'Hold on, General Yates would like to speak with you.' When he came on, I said, 'General, I've been looking forward to meeting you.' He said, 'Good. Get your ass up here.'

"So I went over to see him. He asked me, 'Where did you get this story?' I said, 'You know I can't tell you that.' Then I asked him, 'Is it correct?' He said, 'You know I can't tell you that.'"

King laughs. "Turns out that General Yates had gotten his butt chewed out about my story. The people in Washington, I guess, wanted to know how in the world that story wound up in the paper. So we made an agreement. Yates said, 'If you've got a story that's a bomb like this one, please call me after you write it. I won't tell you not to publish it, but just let me know so I can warn my superiors.' I did that, and we had a great relationship after that."

In 1960, King was offered the job as NASA's chief public information officer at Cape Canaveral. Wernher von Braun and his fabulous rocket team had moved over to NASA, so King knew the United States was about to attempt something big. "And the space program had really affected me by then," he says. "It was fascinating, and I wanted to be a part of it. Plus, I certainly thought I had the credentials for the job after covering the beat for two years.

"My first challenge was preparing for the onslaught of media, once the manned launches started. We had to build our whole press site. We had to work out arrangements to get a television feed for the press. This may all sound real easy now, but we developed from scratch."

He also had to educate media members assigned to cover spaceflights. "It had never been done, so nobody knew much about it," he says. "But we had a main cadre of people who started writing about space right after Sputnik went up. Those people knew what they were doing. They knew the program and did their homework. Others, you'd have to take them aside and walk them through it."

But perhaps the thing he worked on most was making NASA officials at the Cape—including the Original 7 Mercury astronauts—realize the importance of positive press.

"It was hard because they were picking up a newspaper every day, and usually, there was a negative story in there about how bad we were screwing up things," King says. "Learning to launch rockets took time, and we had our fair share of negative stories. So I had to do about as much public relations work *inside* as I did outside NASA. The astronauts would ask me, 'Why should I help them? All they do is blast us!' I don't want to blow this out of proportion, because all of them weren't like that. They just didn't see the need of being interviewed by the media. They were test pilots and weren't used to a lot of attention.

"But we wanted an open program from the start. With [NASA chief James] Webb on the administrative side, and then when Julian [Scheer] arrived on the public affairs side, that was just a fantastic combination. James Webb *made* this program—the influence he had on the Kennedy administration, as far as pushing things through. Those were considered to be desperate days. By the time Alan Shepard had gone up, two Russians and a dog had already orbited the earth. People can't understand it now—the competition involved in the space race—and we were behind."

Still, even when the Russians clearly had the early upper hand, King says he never believed they would beat the United States to the moon. "They took the lead because they were ahead of us in the [Intercontinental Ballistic Missile] races, and that's what put [the first cosmonaut] Yuri Gagarin up in orbit," he says. "But we were ahead in what I call 'miniaturization.' The Russians were like a . . . How can I put this? . . . They believed in the philosophy of using a big truck to just shuffle people and things up there. And I give them all the credit in the world. But in 1972, I had the chance to visit Moscow when we were preparing for the Apollo-Soyuz mission. When I saw their facilities and the way they operated—and I'm certainly no trained technical observer—you could just tell they didn't have the preciseness and capabilities that we had."

As the American space program gained momentum, so did Cape Canaveral. In a few months, it went from an isolated place to a community of about 6,000 residents.

"The Cape was one of the most unique small towns in the country," King says. "All these people coming in here were coming out of the top engineering programs. They were well educated, obviously, because of the job required. They were fairly well paid as a result of that. You throw people together like that, and it becomes a special place. We started building schools, churches. There was nothing here. I happened to be Catholic, and the nearest Catholic church was seven or eight miles across the causeway [in Cocoa]. But we soon had everything we needed right here."

Soon, the space program could say the same thing.

Von Braun's team had been working at the Marshall Spaceflight Center, in Huntsville, Alabama, developing a rocket powerful enough to carry men to the moon. It became known as the Saturn V, and its first test launch occurred on November 9, 1967. Neither King, nor the contingent of media on hand to cover the unmanned mission, was ready for the experience of being in the presence of such raw power—7.5 million pounds of thrust, or fifty times more rocket than Alan Shepard rode into space. It was a controlled combination bomb and earthquake.

"We were in the [launch] blockhouse, and when the engines went to full power, all the dust from the ceiling came pouring down on us," King recalls, laughing. "The windows shook. The ground shook. [CBS News anchor] Walter Cronkite was in his press trailer near the launch site, and he was knocked right out of his chair. Everybody says this now, and it's hard to explain to people, but you had to *be* there. Every one of those Saturn V rockets made the ground shake."

A Saturn V launch, and the hours leading up to it, was tedious and complicated. King's background was in reporting and public relations, not rocket science. But since he was the public's guide through countdown and launch, he wanted to understand it the best he could. His tutor was Clyde Netherton, who helped put together the countdown procedure book for the Saturn V.

"Basically, Clyde was orchestrating the whole countdown. Every hour, every minute," King says. "It was written almost entirely in acronyms, which I didn't understand. But I had the opportunity to sit down with Clyde, and he would go over the whole procedure. He had marvelous patience with me, and we spent many hours together. I'd say, 'OK, Clyde, we're twenty hours before launch. What does this mean?' And he literally took me right down to the final seconds of a launch. He taught me the whole thing.

"I did the same thing with Skip Chauvin. He was the spacecraft test conductor, and he particularly helped me understand the final minutes and seconds, what to expect, what to look for."

During a countdown, King sat on the back row of the Launch Control Center next to Norm Carlton, the launch vehicle test conductor. "I wanted to be right there, hearing what Norm was saying," King explains, "because if there was a problem, he would know. Plus, I was looking at the different status boards as the computer clicked off various activities."

During Apollo, the three television networks—CBS, NBC, and ABC—requested King go on the air with the countdown at T minus three minutes, thirty seconds. He didn't have a script, but thanks to Netherton and Chauvin, he knew the countdown by heart and the importance of each phase. "I did have some notes scribbled down, in case there was a situation where the astronauts had to abort and use the escape tower to blow the command module safely away from the rocket," he says. "I always kept them at my right hand side in case I needed them, so I could explain what was happening."

The Apollo 11 launch, early on a Thursday morning in July, is the one King is best known for, simply because various cable networks air the videotape whenever an anniversary of the mission rolls around. Usually, it picks up the broadcast with less than a minute before liftoff. But throughout the morning, King had given the media and general public insightful updates. An example:

"*This is Apollo/Saturn Launch Control at one hour, seven minutes, twenty-five seconds and counting, countdown still proceeding satisfactorily.*

For those people who would like to synchronize their watches in relation to the count, we'll synchronize on twenty-six minutes past the hour, which is now about sixty-five seconds away. We'll count down the last five seconds to twenty-six minutes past the hour. We're now one minute away from twenty-six minutes past the hour. In the meantime, we do have information from the Civil Defense Agency in the area. The estimate is more than a million persons are in the immediate area, in Brevard County, to watch the launch. Now forty seconds away from twenty-six minutes past the hour. Civil Defense Agency reports further that there is extensive heavy traffic, a number of traffic jams, particularly in the area of Titusville and the U.S. 1 and Route 50. Countdown still progressing satisfactorily. Fifteen seconds away from twenty-six minutes. Five, four, three, two, one, Mark. 8:26 a.m. Eastern Daylight Time. We're now one hour, five minutes, fifty-five seconds and counting, as it was announced at that point."

And this report at 8:41 EDT: "*We have passed the fifty-one-minute mark in our countdown. We're now T minus fifty minutes, fifty-one seconds and counting. Apollo 11 countdown is still GO at this time. All elements reporting ready at this point in the countdown. The spacecraft—correction—the test supervisor Bill Schick has advised all hands here in the control center and spacecraft checkout people that in about thirty seconds, that big swing arm that has been attached to the spacecraft, up to now, will be moved back to a parked position, some five feet away from the spacecraft. We alert the astronauts because there is a little jolt when this arm is moved away. It will remain in that position, some five feet away from the spacecraft, until the five-minute mark in the count, when it's completely pulled away to its retracted position. It's coming up now. In five seconds, the swing arm will come back. Mark.*

"*The swing arm now coming back from the spacecraft. Countdown proceeding satisfactorily. We've completed our telemetry checks with the launch vehicle, and at this point, with the swing arm back, we arm the pyrotechnics so that escape tower atop the astronauts, atop their spacecraft, could be used if a catastrophic condition was going to occur under them with the launch vehicle, from this point on down in the countdown. We have the high-speed elevator located at the 320-foot level, in the event the astronauts have to get out in a hurry. This is a special precaution. One of the members of the support team for Apollo 11, Astronaut Bill Pogue, is here in the firing room. He acts as capsule communication during the countdown. His call sign is Stoney. He controls that elevator. He now has it locked at the 320-foot level. These are special precautions for safety purposes during the final phase of the count. Now coming up on the forty-nine minute in the countdown.*"

King's voice wasn't as relaxed as the countdown melted away.

"*Thirty seconds and counting*," King said. "*The astronauts report it feels good. T minus twenty-five seconds.*"

"I've been kidded about that 'feels good' statement on a few occasions," King says. "I have a buddy over at the Johnson Space Center in Houston who really gives me the business about it. But Neil Armstrong said that. I heard it and reported it. I think he meant just what he said. It had been a tremendously flawless countdown; we only had a couple of minor problems. Everything clicked down perfectly. He was reacting to the fact that things had gone so well and they felt comfortable—as comfortable, I guess, as anybody can feel right before they're about to be launched atop a Saturn V."

"*Twenty seconds and counting.*"

On television, the Saturn rocket—taller than the Statue of Liberty—looked like a toy that might have been propped up by some first grader the day before. There were no signs of the controlled chaos that was about to occur.

"*T minus 15 seconds. Guidance is internal.*"

"Before that, the guidance system had been on ground power so everyone could make sure things were working properly," King explains. "At that point, the guidance switches over, and the spacecraft itself is in complete control."

"*Twelve, eleven, ten, nine, ignition sequence start.*"

"At 8.6 seconds, all five engines begin to build up to roughly about a million pounds of thrust each," King explains. "So during the sequence, the computer is sampling each of those engines to make sure they're building up properly. If for any reason one of those engines was not operating as it should, the computer would shut down the sequence. This is a liquid powered engine, and so there are pumps in there, just like starting an automobile it just builds up."

"*Six, five, four. . . .*"

As the countdown reached zero, King reported, "*All engines running.*" All hell was breaking loose on Pad 39A. "At that point, the computer is taking one last sample of the engines," he says. "So that's a crucial instant."

He reported the rocket left the ground at thirty-two minutes past the hour, "because the news media always wanted to know the exact time of the launch." After he reported "*tower clear*," ten seconds after liftoff, the mission was turned over to the crew at the Johnson Space Center in Houston. King was then free to turn toward the huge window at the back of the Launch Control Center and watch Apollo 11 climb majestically through a few, puffy white clouds.

Apollo 11, with the Saturn V engines creating 7.5 million pounds of thrust, slowly rises from Launch Pad 39A at Kennedy Space Center. [Courtesy NASA]

Within seventy seconds, it was four miles high and . . . gone. Just a bright dot in the sky. Human beings—Americans—on their way to a world previously untouched by anything with a soul.

Other moments were burned into King's memory during launches, such as Apollo 12 being struck by lightning. Twice. "We saw it hit," King says. One of his closest astronaut friends, Pete Conrad, was aboard that flight. They lived on the same street in the early days of the program, when the astronauts worked out of the Cape. King's sons played with Conrad's sons. "I knew all the astronauts fairly well," he says, "but you have to remember, they were later based in Houston. We weren't close buddies or anything, but we knew each other." When Conrad died in a motorcycle crash in 1999, King's son Chip—who flew the longest F-14 combat mission in history, an 1,800-mile roundtrip assault on Afghanistan one month after the 9/11 terrorists attacks—took part in the flyover during the astronaut's funeral.

But some of King's fondest days at the Cape didn't occur during Apollo. "I saw a story not long ago about the launch of Tyro, our first weather satellite," he says. "I did that launch. That was in 1960, when it was impossible for humans to look down on the earth. It was a remarkable step forward. A lot of those things have been forgotten by the public, but not by me."

King also isn't one to gloat about "the good ol' days of Apollo." After leaving NASA in 1975 and working with the Energy Department in Washington for a couple of years, and with Occidental Petroleum for fifteen, he and his wife Evelyn returned to Cape Canaveral in 1997. King took a job in media operations with the United Space Alliance and quickly found out the people working the shuttle program also took their business seriously—the missions just weren't as sexy as Apollo.

"A lot of what's been said and written about the space program is understandable, about poor morale and different things," he says. "But I will never forget when Neil Armstrong put his first boot print in the moon about two thousand people were laid off. A lot of people's jobs were done, and that was a sad thing. And then we had the energy crisis, the environmental crisis. The Cold War was still going on. Everything wasn't perfect."

All is not perfect in Jack King's world anymore, either.

In March 2005, his wife of thirty-nine years passed away. "My pal, my lover, and my soul mate," he says of Evelyn. They were married in Cocoa Beach on July 17, 1965, and had three children—Chip, Beth, and Billy—and five grandchildren. They loved living at the Cape. The kids never viewed it as the place astronauts began their journeys to the moon. To them, it was where youngsters learned to ride their bikes and catch fireflies in the coastal peace of gloaming.

He's had a good life, King emphasizes. He was part of, perhaps, the greatest achievement of the twentieth century. "Maybe even the millennium," he says. His voice will remain synonymous with the moon launches, "something I can take a little pride in," he says.

"The turnaround this program made after losing Gus Grissom, Ed White, and Roger Chaffee in the Apollo I fire on the pad [in January 1967] was simply amazing," King says. "I don't think anyone expected us to recover that quickly."

With barely more than five months to spare, NASA fulfilled President Kennedy's 1961 challenge—before an American had ever been into earth orbit—to land a man on the moon and return him safely.

"In a cornball manner," King says, "Kennedy had his Camelot. And I guess we had ours with Apollo."

PART IV

Marriage, Missions, and Moon Cars

11 JOAN ROOSA

Wife of Apollo 14 Astronaut Stuart Roosa

Going to the moon brought more than fame to the twenty-four astronauts who flew there.

"They had groupies too, just like rock stars and movie stars," says Joan Roosa, whose husband Stuart was command module pilot on Apollo 14. "They were world heroes, and there were women—especially down at the Cape—who chased them. I was at a party one night in Houston [when] a woman standing behind me, who had no idea who I was, said, 'I've slept with every astronaut who has been to the moon.' Well, my husband had already been there. So I turned and said, 'Pardon me, but I don't think so.' Stuart was monogamous, and I had more than one astronaut tell me how much they admired him for his faithfulness to me.

"But that's the way it was back then. The astronauts were a big deal, and women made them a target."

For all its glory, the Apollo program put a tremendous strain on the families involved. Only two of the twenty-four moon voyagers were bachelors at the time—Jack Swigert and Jack Schmitt. Twelve of the other twenty-two went through divorces.

But it wasn't always the fault of groupies or carousing pilots.

"Everybody wants to blame the astronauts," Joan Roosa says. "But I think more divorces were started by the wives than the astronauts. That may shock some people, but it's the truth. I know for a fact Neil Armstrong never wanted a divorce from Jan. He was served with divorce papers three times, and he kept throwing them away. [Neil and Jan Armstrong divorced in 1990, and he remarried in 1994.]

JOAN ROOSA

Taken on the anniversary of the Apollo 11 moon landing, July 20, by Dot Cunningham (wife of Walt Cunningham, Apollo astronaut). Stuart Roosa passed away a few months later. [Courtesy Joan Roosa]

"It was tough for everybody. We called the whole moon thing 'the program,' and everyone involved had to be dedicated to it. The program dominated every waking moment. And it was particularly rough on the wives. All week long, while your husband [was] preparing for his mission, you [were] running the household. Literally, doing everything. Then on Friday night, in walks Daddy. That meant fun time. All week long, we were telling the kids, 'No, you can't do that.' Then Daddy shows up and he is ready to do whatever the kids want to do. That caused problems, but it was something the wives had to keep to themselves. We smiled for the photographers. We said what we were supposed to say. There were downsides to it all, believe me. The wives had to sacrifice a lot.

"But, personally, I was devoted toward going to the moon, just as much as Stuart. It was terribly exciting that man was going to leave the confines of earth. I think, as a wife, you had to understand the big picture and be devoted to it in order to stay together like we did."

Roosa, whose husband died in 1994 due to complications from pancreatitis, may have had an advantage over the other wives. "I knew from the start what Stuart wanted to do," she says. "He was a test pilot when I met him, and I came to realize very quickly that a career like that is a day-to-day situation. A lot of the wives met their husbands in college, before they ever knew what they wanted to do. When they later became test pilots or astronauts, maybe there was some resentment because of all the sacrifices they had to make. They didn't picture their lives going in that direction."

Joan Roosa didn't have much reason to complain about Stuart's workload as an astronaut. She was the one who insisted he apply for the job.

Joan Barrett Roosa, a southern girl, went to elementary school with Elvis Presley in Tupelo, Mississippi. Stuart Allen Roosa was a country boy from Claremore, Oklahoma, who knew from the time he was nine years old that he wanted to fly fast airplanes.

They met in 1956 at the Langley Air Force Base Officer's Club in Hampton, Virginia. Joan, along with several of her friends who had just graduated from Mississippi College for Women, was teaching school in Hampton. Stuart was a test pilot at Langley.

"One of the teachers I lived with had a cousin who was a pilot at Langley," Roosa recalls. "It was her birthday, and her mother flew up from Tupelo to take us out to dinner, and the girl's cousin arranged for us to go to the Officer's Club. Stuart had been to a party for his squadron, and he had decided to stop by and have a drink at the bar. I walked across the room—at least this is the story I heard him tell a thousand times—and he told the bartender, 'That girl who just walked by here is going to be my wife.' We'd never even met! But he was very persistent, when he wanted to be, and he finagled his way over to where we were. We danced and talked.

"He asked me on our third date to marry him. I told him I just couldn't do that. I lived with other teachers, and there were always a lot of people in the house. One day he was over, and he just blurted out, 'Joan said she'd marry me.' Well, being a nice southern lady I didn't want to correct him in front of all those people. And the more I thought about it, [the more] I realized, I really [did] want to marry this man. He was so romantic with his silver jet."

They were married in 1957. Seven years later, Stuart was flying at Edwards Air Force Base in California when NASA put out a call for more astronauts. Joan mentioned it to him, but Stuart balked. "He said he had the job he'd always wanted," Joan recalls. "But I wouldn't let it go. I said, 'Why don't you just *try* it?' So he did. Thousands applied, and NASA kept cutting and cutting and cutting. One day he came in and said that a pilot friend of his, who had also applied for the astronaut corps, Joe Engle, had gotten a letter asking him to continue. He said, 'I didn't get a letter, so I guess that's that.' I said, 'No, I don't think so,' and handed him his letter. It had come to the house instead of his office, and I think he really realized how much becoming an astronaut meant to him at that point."

The finalists were invited to Houston for an interview exam with several NASA administrators, including Deke Slayton, chief of the astronaut office. He was *the man*, the one person—sometimes influenced by America's first space traveler, Alan Shepard—who made crew assignments. Stuart Roosa had studied hard.

His interview session was barely under way when Slayton received a telephone call and left the room. He never returned. When Stuart called home that night, Joan could detect the disappointment in his voice. "He said he knew it was over when Deke didn't stick around," Joan says. "But it turned out that Deke was very good friends with a commander at Stuart's test pilot school, and he had told Deke, 'You want Roosa at NASA.' So Deke had already made up his mind that he was going to pick Stuart." She laughs. "All that stress for nothing."

Joan was thrilled—not only for Stuart, but for herself. His selection meant they would be moving to Houston, a happening town with the newly built Astrodome—the Eighth Wonder of the World—and a space program that was capturing the world's attention. "Edwards was on the high, dry Mojave Desert, and there wasn't a town within thirty miles," she says. "I wanted out of there."

Everyone at the Manned Spacecraft Center stepped lightly around Alan Shepard, whose persona was much larger than his five-foot, nine-inch frame. He was so intimidating, the story goes, that his secretary kept two pictures of Shepard—one of him smiling, the other displaying a disgusted frown. For visitors' sake, she would display the picture that reflected Shepard's mood of the moment.

"Stuart was scared to death of him," Joan says. "He used to say that whenever he saw Alan Shepard coming down the hall, [Stuart] would duck into the next office, no matter whose it was."

So he was more than a little nervous when Shepard summoned him and fellow rookie astronaut Edgar Mitchell to his office one afternoon in the summer of 1968. "Stuart said he was literally quaking in his boots," Joan says. "But Alan Shepard said, 'I want to ask y'all something? Do you mind flying with an old retread like me?' Stuart was so taken aback because he had never even served on a backup crew. He looked at Alan Shepard and said, 'You mean on the backup crew?' Shepard gave him that icy stare of his and said, 'I never mentioned anything about a *backup* crew. I said crew.'"

At that moment, Stuart Roosa realized he was going to the moon. Not until years later would he learn that Shepard had handpicked his crew. "Shepard said at a banquet one night that he selected Ed Mitchell because he knew more about electronics than anyone he'd ever known," Joan says. "And he said he picked Stuart because he was the best stick-and-rudder man he'd ever seen, which was an incredible compliment coming from Shepard. He could've flown with anyone he wanted. Plus, he and Mitchell were both Navy guys; Stuart was from the Air Force, and there was always kind of a rivalry between the two over who had the best pilots."

Shepard, Roosa, and Mitchell originally were scheduled for Apollo 13. But because Shepard had been grounded for years with inner-ear trouble, NASA brass figured a few extra weeks in the simulator would do him good. So they were moved to Apollo 14, and the original Apollo 14 crew moved up to 13. Little did anyone know how history would be affected by that move.

Apollo 13, crippled by an explosion in the service module, didn't land on the moon and was nearly lost in space. Stuart Roosa proved invaluable in the simulator, as ground controllers wrote new, intricate procedures to help get Jim Lovell, Fred Haise, and Jack Swigert back to earth. Meanwhile, Joan Roosa was performing her own form of ground support. Whenever an emergency at NASA arose, it was standard procedure for the wives of each astronaut class to help each other in any way possible—answering the phone, keeping the press at bay, tending to the children. Even though Joan was closer to Marilyn Lovell and felt like she should be with her, their husbands had joined the astronaut corps at different times—Lovell in 1962, Roosa in 1966. So Joan wound up sitting with Haise's wife, Mary, who was eight-months pregnant and wondering if her husband would ever see their fourth child. "It was tough," Joan says. "You just try to remain positive and let them know they aren't alone. . . . There's really not much else you can do." Privately, she admits, she thanked God the crews had been switched and that it wasn't Stuart who was stuck up there. "I think that's only being human," she says.

Apollo 13's successful return was one of NASA's proudest moments because of the ingenuity shown by those in mission control, and also the three astronauts who never allowed panic to rob them of their trouble-shooting skills. They saved a mission that appeared hopeless. But it was clearly a huge setback. Suddenly, a program that was launching a moon mission every four months was grounded for nearly ten. And unbeknownst to Roosa and the other astronauts, President Nixon was getting heavy pressure from his political advisers to immediately end the moon program. We had beaten the Soviets, collected 128 pounds of samples during two landings, and every astronaut who was launched had come back alive. Quit while we're ahead, they urged him.

Nixon later shared this with Stuart Roosa. He told the astronaut, "I went against every piece of advice I got and decided to carry on with Apollo." The only thing Stuart could think to say in return was, "Mr. President, I'm sure glad you did."

No one in the astronaut corps understood the hazards of the job more than Roosa. He had lost buddies who flew combat missions in Vietnam. He knew about the crashes of T-38 jets that claimed the lives of astronauts Charles Bassett, Elliott See, and Theodore Freeman. And in January 1967, Roosa was the capsule communicator during the fatal Apollo 1 launch simulation at Kennedy Space Center when astronauts Gus Grissom, Ed White, and Roger Chaffee died in a fire. Roosa spoke the last words they ever heard, listened helplessly as they screamed for somebody to get them out, and experienced the sickening smell of burned flesh. He had been especially close to White, who lived in the same neighborhood as the Roosas.

"But Stuart was a pilot—first, last, and always," Joan says. "He loved flying so much that he was willing to take the risks that went with it. That's one reason he wanted to be a command module pilot, which meant he would stay in orbit while the other two [astronauts] landed on the surface. He wanted to *fly* that spacecraft.

"It was hard when I was asked by reporters if I was scared about Stuart going to the moon. The press wanted a yes or no answer, and there was no way to answer that question with a yes or a no. Was I apprehensive about it? Naturally. But would I keep him off the flight if it were in my power? No, because it had been his dream. He had worked hard for it, and the space program needed a boost after what happened on Apollo 13."

NASA was at a serious crossroads, and it was up to Apollo 14 to keep the program going, at least a while longer.

Joan Roosa paid her own way to Cape Kennedy for the launch. She had seen Stuart only a couple of times during the previous three weeks. When they weren't training, the astronauts were holed up in a Florida

The Apollo 14 crew consisted of Ed Mitchell, Alan Shepard, and Stuart Roosa. Roosa was content to remain in the command module and "solo" around the moon while Shepard and Mitchell flew to the surface. [Courtesy NASA]

beach house, away from as many germ-carrying humans as possible. The wives were among the few allowed to visit.

On the eve of the launch, the wives were taken to a site near Pad 39A for a look at the monstrous Saturn V their husbands would ride into orbit the next morning. Joan was stunned by what she saw.

"It almost looked alive . . . like some gentle beast," she remembers. "It was thirty-six stories tall, and the little capsule where the astronauts sat was no wider than my couch. They were filling it with fuel, which made the rocket so cold, ice formed on the outside of it. And they had these large lights shining on it from different directions. It sparkled like a diamond.

"To know your husband was going to sit atop that thing and travel so far away. . . . It was something you felt down in your stomach."

When she arrived back home in Houston, a few hours after a perfect launch, her front lawn was filled with writers, broadcasters, and photographers.

"I tried to be as nice and cooperative as possible," she says. "I kept a pot of coffee out in the garage, with cups and sugar and cream. But Sue

Bean gave me the best advice I ever received on handling the press during the flight—she learned the hard way when [her husband] Alan went up on Apollo 12. She said, 'The photographers have to have a picture for their newspapers the next day of the wives. Get yourself looking nice, go outside in the morning, let them take your picture, answer any questions anybody has, and then a lot of 'em will drift off and leave you alone the rest of the day.' And that was true."

She also had prepared their four children—Christopher, eleven; Jack, ten; Stuart, eight; and Rosemary, seven—for the media onslaught. "I told them, 'You're going to have nine days of being treated like a big deal. Photographers are going to follow your every move. And then Daddy's going to get back, and he's going to become just another guy.' I poured that into their brains and never let it enter their minds that he *wasn't* coming back. I said, 'Yes, you will be coddled to by the press. But remember it's going to be over, and it's going to be another daddy down the street who's going to the moon, and their kids will be the big deal then.'

"One of Buzz Aldrin's sons told me once that he wished someone had told him that before his daddy went up on Apollo 11. He said that the whole time Buzz and Neil were on the moon—and let's not forget Mike Collins, the command module pilot—that he thought the fame was going to last him the rest of his life."

Despite all the coaching he had received from his mother, Christopher Roosa discovered just how much his life had changed on the second day of the mission. "I did something back then that doctors would have a fit over now—fixed him a fried egg on toast," Joan says, with a laugh. "Well, he goes out to the end of the driveway to get the paper that morning, while I'm cooking. It's still dark. And all of a sudden he comes running back inside and just looked horrified. He said, 'Mama, I never made it to the paper. A whole bunch of flashbulbs went off in my face, and it scared me so bad, I just turned around and ran!'"

Anything about an astronaut or his family was big news during a mission—even an eleven-year-old, still dressed in his pajamas, retrieving the *Houston Post.*

"People ask me the scariest moment of Stuart's mission, and they expect me to say the launch or when they fired the engines to return from the moon," Joan says. "But no, the scariest moment for me was having our children interviewed live by [CBS news anchor] Walter Cronkite because you never really know what's going to come out of a child's mouth. I knew they were going to ask all the children questions, and I had sort of warned them about saying the wrong thing.

"Walter Cronkite asked Christopher, 'Do you want to be an astronaut?' And Christopher told him, 'No, I think I want to be a doctor.' He had just had his appendix removed, and he was very impressed with his doctor. He asked John the same question, and John said, 'I think so.' He asked Stuart, and he said, 'Oh, yes sir. That would be great.' [Cronkite] finally got around to Rosemary, and he said, 'What do you think of having your daddy flying around the moon?' She sort of looked at me and said, 'It's *greaaaaaaaat*'—as if she were saying exactly what I had told her to say. But, again, that was all part of the program, smiling for the cameras, saying the right things."

Apollo 14 was not without its tense moments. Because of a faulty docking mechanism, it took six tries for Stuart Roosa to link the command module with the lunar module. As Shepard and Mitchell were preparing for their flight to the lunar surface, the LM's guidance computer wouldn't cooperate. They were within a few minutes of having the mission scrubbed before Don Eyles, a young computer whiz at MIT, quickly wrote a program to fix the problem. And as the LM descended to 22,000 feet, the landing radar went awry.

But eventually, after a tough day in space, Shepard and Mitchell landed in the Fra Mauro highlands. And up in the command module, *Kitty Hawk*, Roosa found himself alone—the fourth human to solo around the moon.

Unlike some of the other moon voyagers, Stuart never minded sharing his deep feelings about the journey, especially with his family. On many occasions, he told Joan and the children about the loneliness one feels on the backside of the moon, being out of touch with all humanity and in darkness understood only by those who flew there. "He said it was either really bright or really dark, and nothing in between," Joan says. "Like turning a light on and off."

Perhaps his favorite story was his experience just prior to Lunar Orbit Insertion. A disc jockey in Houston, and a friend of Stuart's, had arranged for the astronauts to take along a specially recorded cassette tape of music by artists such as Buck Owens and Sonny James. As Stuart spotted a large crescent moon filling one of *Kitty Hawk's* windows, the hymn *How Great Thou Art* began playing on the tape player, as if scripted. It had been one of his favorites since childhood.

> O Lord my God / When I in awesome wonder / Consider all the works thy hands have made / I see the stars, I hear the rolling thunder / Thy power throughout the universe displayed.

"I [heard] people say that Stuart had to go to the moon to find religion, and that used to really irritate me," Joan says. "It was a ridiculous statement. Stuart was religious when I met him. And he took his faith seriously.

"He was raised in the Baptist church, and I'm Catholic. When I was teaching down in Virginia, and Stuart and I were dating, he used to fly early in the morning, and then . . . pick me up when I got off around four o'clock. One day I said, 'You picking me up today?' And he said, 'No,' which totally surprised me; he usually took every chance he could to see me. But then he said, 'Joan, I've started taking lessons in Catholicism. I've decided I want to be a Catholic. I'm not doing this just to please you. I feel I was born to be a Catholic.' He became a Catholic, and we were married in a Catholic church."

Apollo 14 is perhaps the most unsung of the lunar missions. It got the program back on track. Shepard and Mitchell retrieved ninety-four pounds of lunar soil and rocks, and Roosa's photography helped in the selection of future landing sites, when geology would become even more of an emphasis.

It didn't take long for Joan Roosa to fully understand that we look at people differently once they've flown to the moon. "They were in quarantine for three weeks, and we could only talk to [them] through glass," she says and then pauses. "You know, they never really told us what they would do with them if they [came] back with some strange disease." She laughs. "Maybe *they* didn't even know.

"The day they got out, we were paraded around at the Houston Rodeo, which is a really big deal in Texas, and then we were on our way to Washington for dinner with President Nixon."

Joan Roosa noticed her assigned spot at the table was beside Nixon. "As I went to sit down, President Nixon held my chair for me," she remembers. "And I got almost down in sort of a half squat, and I just froze. All of a sudden it hit me, 'You are Joan Barrett from Tupelo, Mississippi and the President of the United States is holding your chair for you in the White House!' I'm sure the president was wondering 'What's the hold up?' But it was really something I'd never even dreamed of."

Soon, Stuart Roosa was back to the grind of training. He was on the backup crews for the final two Apollo missions, 16 and 17. He stuck around for the beginning of the shuttle program, but he soon became bored. Joan says, "He'd come home complaining, 'We're talking about things we worked out ten years ago. Why are we going back over them?' He realized it was time to find something else to do."

Stuart retired from NASA in 1976—but not before he and Joan traveled the world as goodwill ambassadors. They went to France, Saudi

On a speaking tour after the Apollo 14 mission. Louise Mitchell (wife of Ed Mitchell) on the left; Joan Roosa (wife of Stuart) in the center (purse in front was hand-painted by Joan); Louise Shepard (wife of Alan Shepard) on the right. [Courtesy Joan Roosa]

Arabia, South America, and Afghanistan. But the trip she remembers most was to Nepal, a country of 22 million, located between India and China. It was part of the People-to-People program; Stuart helped deliver all sorts of athletic equipment to poor children there.

"I went down to the beauty shop in the hotel we were staying at to get my hair done," Joan says, "and the women in there were buzzing like bees. I had no idea they were talking about me. Finally, one of them said, 'Are you married to the astronaut?' I said, 'Yes, that's right.' The woman said, 'You realize that makes you a goddess.' Well, I had a hard time not laughing out loud right then, but I couldn't wait to get back upstairs and tell Stuart he was married to a goddess."

But the Nepalese were also serious. As Stuart and Joan traveled to a function that evening, children holding candles knelt beside the road in Stuart's honor.

When Stuart spoke at a school the next day, he was asked who he saw as he was circling the moon. "Nobody," he answered. The children and adults in the room looked at one another in disbelief. "Who did you see when you were on the *dark* side of the moon?" they inquired. "Nobody," he replied again. "There was nobody on the moon."

The Roosas were furious when they later found out that many Nepalese believed the moon is one's final destination after death, and they were hoping to hear that Stuart had seen or talked with some of their ancestors.

"Somebody should have briefed Stuart on it before we went there, and I blame the State Department for that," Joan says. "Stuart said, 'At least I could've been diplomatic about it. But as it was, I basically told them there was no heaven.'"

When Stuart left NASA, U.S. Industries' International Operations hired him to open an office in Europe, and he was free to select the site. "Being a history major, I picked Greece," Joan says. "I'd studied a lot about it, but had never been there. But that turned out to be a nightmare. The bureaucracy there was awful. It was almost impossible to set up an office; it was tough just getting a phone."

After just a year, the Roosas came back to the United States, and Stuart began working with a company trying to develop an electric car. That didn't last long. He resigned because of disagreements about how the business was being run and spent the next four years working in commercial real estate.

Then, in 1991, Coors called offering him the opportunity to open one of its first distributorships east of the Mississippi River. Shepard and Apollo 16 astronaut Charlie Duke had already enjoyed success with Coors. And the challenge of building a business from the ground up appealed to Stuart.

"They said we could go anywhere in Mississippi we wanted, so I picked the Gulf Coast because I had always liked it there," Joan says. The Roosas moved to Gulfport, where they bought a tiny house near the beach. They went through all their savings and took out loans to get things rolling.

"Everybody thought astronauts made a lot of money," Joan says. "But Stuart was paid the same as any other Air Force Major when he was flying. I used to say we lived in the best and worst possible worlds. We were invited to everything in Houston. Trying to run a household and stay dressed properly for all that stuff; . . . it wasn't an easy juggle.

"And opening the Coors distributorship took a lot of money. We had to get a warehouse and build a big cooler in it. We had to buy trucks, hire drivers, hire supervisors, put together an office staff. And then we had to pay them."

After only one year, the Roosas were debt-free and making a lot more money than an astronaut. They moved into a larger house on a golf course. Late most afternoons, they would go outside. Joan would have a glass of wine. Stuart would drink a beer and play four or five holes of golf. It was an entirely different world than they had known two decades earlier, and they loved it. Stuart also got into big-game hunting all over the world, and Joan accompanied him on most of his trips.

"I think a lot of the astronauts suffered after coming back from the moon," Joan says. "They'd say, 'Oh, the moon didn't change me,' but you could tell it kind of did. These were men who had been to the mountaintop of their profession. What [could] they do that [could] compare to going to the moon? In Stuart's case, I think the safaris we went on showed him that there were wonderful, exciting places right here on earth that he had never seen. And he really got a charge out of building the [Coors] business. Those things together helped Stuart adjust to life after NASA."

Stuart turned the business over to daughter Rosemary in 1988. "All the boys were in the military at the time, and none of them seemed really interested in the business," Joan says. "But Rosemary was ready for this." Rosemary, a redhead just like her daddy, was a graduate of Harvard Business School. She ran the distributorship for ten years, and "[she] ran it very, very well," Joan says, before selling it in 1998.

As a semi-retired couple, Stuart and Joan Roosa began each morning by reading two newspapers—the Jackson (Mississippi) *Clarion-Ledger* and the Biloxi *Sun Herald*. Then they'd talk about the stories of the day.

"I saw in the paper one morning [that] somebody I knew had died," Joan recalls. "And for some reason, I started thinking about things. . . . I had battles with high blood pressure, and I always thought that I'd die before Stuart. So I told him, 'Stuart, if something happens to me, I want you to get married again. I don't want you to be by yourself. I wouldn't be jealous. We've had a good marriage, and I'd rather see you happy with someone else than be lonesome.' He said, 'I don't think I could do that.'

"So I said, 'Well, just promise me one thing—that you won't commit suicide if I die.' He sat there for a couple of minutes and said, 'I can't even promise you that.'"

In 1994, the Roosas traveled to Washington to spend Thanksgiving with Christopher and his family. Stuart began having excruciating abdominal pains, and a doctor Christopher knew examined him. He immediately admitted Stuart into the hospital. The diagnosis was pancreatitis.

"I went on and cooked dinner because I figured he'd only be in there a couple of days," Joan says. "The doctor was even talking about having Stuart speak to his son's class at school. Then, all of a sudden, he got much worse."

Twelve days later, Stuart Roosa died of pneumonia, a result of the pancreatitis. He was sixty-one.

Years before, at Joan's urging, Stuart had filled out the paper work to be buried at Arlington National Cemetery. "He died on a Monday, and Christopher said he wanted his daddy buried on Thursday," Joan says. "The man at the funeral home said that was impossible. Christopher said, 'Yes, it's possible. You just handle your end, and we'll take care of the rest.'"

Sadly, the man had no idea who Stuart Roosa was. "His name was never well known like Neil [Armstrong's]," Joan says. But on that Thursday, Stuart Allen Roosa was buried with a full military service, including a flyover that shut down Reagan International Airport for fifteen minutes. Alan Shepard, the guy Roosa used to be so afraid of, delivered the eulogy.

"Stuart's death was a shock to all of us. It happened so suddenly," Joan says. "But we had thirty-seven years together. And I think about how fortunate I was to be married to Stuart and do some of the things we did together. Rosemary says all the time, 'Mama, I've had a good life. But you've had a *fabulous* life.' And I can't argue with that."

In February 2005, on the twenty-fourth anniversary of Apollo 14's splashdown, the Roosa family gathered once more at Arlington cemetery. They planted a tree grown from one of approximately 500 seeds Roosa packed inside his personal kit on Apollo 14, a tribute to his early days as a smokejumper with the U.S. Forest Service. "Moon trees" from those seeds—sycamores, Douglas firs, loblolly pines, sweet gums, and redwoods—now stand on lawns from the White House to Japan.

Since the death of her husband, Joan Roosa has lived a quiet life near the beach in Gulfport, just around the corner from her sister and only a few blocks from Rosemary. (Their homes survived Hurricane Katrina in August 2005.) Joan has traveled with her daughter, and another female friend, to places such as China and Jordan. And whenever she sees the moon, she remembers Stuart looking up at it following his flight and saying, "My trip there seems a lifetime ago."

For nearly two years following his death, a steady flow of mail addressed to her husband arrived from autograph-seekers. Finally, as more people learned of his passing, the letters slowed to only a handful a year.

"I'm glad," she says, "because I had to learn to write, 'Deceased. Return to Sender' on the envelope. Do you know how hard that was for me to do?"

12 RODNEY ROSE

Mission Planner

Planning a space mission during the hectic days of Apollo was like having a few thousand world-renowned chefs in the same kitchen. They all packed an ego and their own ideas about how things should be done, and they weren't shy about expressing them.

Rodney Rose, a native of Huntington, England who had been on NASA's wish list since his days of designing aircraft for the British government, specialized in the capsule's ejection system and parachutes during the Gemini program. He heard the constant bickering back and forth among contractors, engineers, scientists, and medical experts. They argued over everything from electronic configurations to what, and how often, astronauts should eat in space. Rose pled his cases with the best of them.

In 1964, as Gemini was about to begin and Apollo loomed on the far horizon, he approached Chris Kraft, director of Flight Operations. "I told him we needed to get everyone on the same sheet of music, as if we were playing a concert," Rose says, his British accent holding strong after forty-six years in Texas. "Everyone was going off every which way, and Chris and I were pretty much on the same wavelength. He put me in charge of these [Flight Operations Panel] meetings and said, 'Do the job and I'll back you.' Pretty soon people found out that's exactly how it was; they didn't buck me too much because they knew Chris would come down on them."

Rose's formal title became Technical Assistant, Assistant Director of Flight Operations. Basically, he was a mission planner who made sure flights maximized every precious second in space without overloading

RODNEY ROSE

A native of England, Rodney Rose helped schedule every minute of the Apollo missions' flight plans. "Time in space was too precious," he says, "but I didn't want them overloading the astronauts, either. It was a fine line to walk, but I loved what I did." [Courtesy Rodney Rose]

the astronauts. He ran up to ten meetings a week. "You have to remember," he says, "we were planning past the next mission. It was the one after, and the one after that. Everyone involved in a mission would come up with a rough outline of a flight, bring it up before the group, and we would chew it, bit by bit, all the way through. I made sure that what they did made sense, and then put Chris' feelings on it—which were mine, also, because we'd talk about it beforehand. And then we'd present the plan to the crews."

Rose chuckles. "Of all people, I remember the doctors weren't too happy about having to come to these meetings. But I said, 'Look, you'll learn something.' So they came to the first one, bitching and grumbling. Afterward, they came up to me and said, 'We learned a whole lot of stuff that we had no idea was going on.'"

It took six *years* to develop a final mission plan for Apollo 11, the first attempt at a lunar landing. The first moonwalk, which lasted two hours, thirty-one minutes, real time, took two years to choreograph. "We literally talked about every step Neil and Buzz would make that night," Rose says.

Rose vividly remembers the last FOP (Flight Operations Plan) meeting just weeks before launch to finalize the Apollo 11 flight plan. Approximately 125 people crammed into a room on the ninth floor of the Manned Spacecraft Center's main building. On the front row was the flight crew—Neil Armstrong, Buzz Aldrin, and Michael Collins.

"They came to nearly every one of the FOPs because that's the only place they could find out what the hell they were going to be asked to do," Rose says. "We went through the whole plan that night with everybody weighing in for the final time. It was quite amazing to sit there and realize what we were putting the final touches on. But as we went along, the crew would nod. And at the end, they all three signed off on it. They said, 'We agree with what you've come up with. It's a good plan.'

"I guess you could say that's the night we sprinkled holy water on the plan that landed two Americans on the moon for the first time."

Rose almost didn't live to see Apollo, or even adulthood.

His mother, Flora Kate Rose, wanted her sons far away from the constant warning sirens and the thunder of German bombs that pummeled their county of Huntingtonshire, a county that supported forty-seven air bases during World War II.

"I had family in Canada," Rose says, "so [my mother] was going to send my brother and me to stay there with [them] until the war was over. But at the last minute, they wouldn't let my brother go. He was eighteen months older, and I'm not sure why they told him no. But my mother said, 'If you both can't go, then neither can go.'"

A German torpedo destroyed the ship they would have been aboard.

"What if?" Rose ponders six decades later. "There are lots of what ifs in my life."

He went to the same grammar school as Oliver Cromwell, the seventeenth-century ruler of England, Scotland, and Ireland. He attended on scholarship. "We were a poor family," he says. "My dad [David] worked as a night watchman, and as a kid, I remember not having any toys to play with—only handmade ones."

Rose later went on to Manchester College of Technology, also on scholarship, and Cranfield College of Aeronautics, earning degrees in advance mathematics and aerodynamics.

"The best thing was that I learned how to design airplanes," he says. "I designed a wind tunnel. Wrote my experimental thesis on low speed characteristics of a delta wing, which was published by the college. For some reason, mathematics came easy to me. I loved it."

He worked a year as a gentleman's apprentice at A. V. Roe (Avro), an aircraft manufacturer in Manchester. In 1951 he took a job at Vickers-Armstrong in Winchester Hants, England, where he soon was head of performance and engines. "Basically, I had a key to the executive washroom, a chauffeur, and all that crap," Rose says. "But I was hiring people for about the same salary as I was getting. They kept saying, 'You'll eventually get yours.' But eventually never came."

In 1957, he ran across a couple of his buddies who had served apprenticeships with him. They were working at Avro in Ontario, Canada. Rose recalls, "They told me that they were in the process of developing a new supersonic delta wing airplane called the Arrow, and they said, 'You should come talk to our brass. We'd love to have you. You've got all that aerodynamic information about deltas.'"

So he interviewed and got a job in design. "I loved it," he says. Rose worked there nearly two years. Two weeks before the Arrow was to make its first flight, the Canadian government chopped the program. "There was so much politics involved," he says. "Avro was set up by the liberal government at the time. Then the first conservative Prime Minister, [John George] Diefenbaker, was elected for the first time in no telling how many years. He hated everything the liberals had done. So Avro had a black mark against it to begin with.

"But I learned a valuable lesson there: You can spend a major part of your life working on a project, really get it going, and then it could very easily get canceled. I never wanted to become part of something that I knew didn't have total commitment again."

Job offers came quickly that April of 1959. An engineering company in Ottawa wanted Rose to help develop a wind tunnel. And NASA, which had been formed only six months earlier, invited him to become part of America's space program. Bob Gilruth, head of the Manned Spacecraft Center, had heard of Rose's work on the Arrow, which involved analog simulators and real-time digital data being sent from the airplane. NASA desperately needed engineers with that kind of experience.

"At the time, it was a tough decision. But then I thought, 'You know, Ottawa's a pretty cold place in the winter. I'm going to NASA,'" he says, laughing. "Can you imagine that? A wind tunnel or the moon. They told me in Ottawa, if the space thing didn't work out, that they'd have a job for me. But I'd say I made the right decision."

When he arrived at NASA, Rose was one of twenty-five foreigners placed in middle-to-high management positions. Twelve were from England, twelve from Scotland, and one from Wales.

"Of course there was a bit of friction," Rose says. "We were experienced people who had been doing things that other people at NASA simply didn't have the experience in. I had fourteen years in the business, and I had four or five young American engineers working under me. But you know what? They accepted us like gangbusters. And I don't think that would've happened anywhere else in the world. Not in Canada, not in England. I think that tells you about the commitment of the people in the space program and the feeling among the workers of, 'Let's get the job done.'"

Rose's biggest problem had to do with his accent.

"I had one gentleman, a good ol' Texan, walk up to me one day and say, 'Rod, I wish you would speak English!'" Rose says with a chuckle. "I said, 'My dear friend, I speak nothing but.'

"We were informed, when we arrived, that we would have to file for citizenship in order to get [security] clearance, and that would take five years. One of the ladies asked if we would also like to take a course in English. I told her I would be happy to offer a lecture in it, and she looked at me in a rather strange way. I'm not sure what language they thought we spoke.

"I learned a bit about the way things were in Houston when I received my citizenship [in 1964]. At the end of the ceremony, the judge said, 'Congratulations, you are now Texans. Oh, and by the way, you are also Americans.'"

Rose tried his best to do things the NASA way.

"But I didn't like NASA's grammar," he says. "I was a stickler for that. The word 'data' for instance. Data is plural, so it should always be 'data are.' But not at NASA.

"I was going to give a presentation to the Space Rescue Symposium in Belgium once, and at that time, NASA had a contractor, supposedly made up of all English majors, to write these speeches for us. This poor girl comes into my office—I still feel sorry for her today, I was pretty arrogant in those days—and I really chewed her up and down. I said, 'I *never* write stuff like that. It's the most un-English grammar imaginable.' She said, 'It's in standard NASA document form.' I said, 'Well, the standard document is wrong!'

"But I was the same way with the engineers under me. They would come into my office with lots of data. I would look at it and say, 'How do you explain this?' They would look at me and say, 'I don't know. That's how it came out of the computer.' So I would tell them, 'Go back, plot it out, and figure out what fundamental law caused it to do that. Then come back and tell me.' They couldn't be guessing. They had to understand the principle involved."

Rose was always one to question the logic of anything he didn't believe made good sense—even if it came from a brilliant mind.

"We had big meetings at Langley [Air Force Base in Hampton, Virginia] and up in Washington, and there were big arguments about the best way to go to the moon," he says. "A lot of people—[rocket guru] Wernher von Braun included—were in favor of a direct approach, with one spacecraft going out, landing, and coming home. Others thought the lunar orbit and rendezvous with two vehicles was the way to go, and I was in favor of that."

Rose admits he was "one of the few" who still held a grudge toward von Braun, simply because he was German. When Rose was around him, he didn't see the man who eventually fled Germany because he wanted no part of Hitler's war efforts. Instead, all Rose could picture were the bombs that had pounded his country, killing thousands and scaring his mother to tears. He knew von Braun's team of engineers had helped develop the rockets that carried those bombs. "But I eventually said, 'what the hell' and buried the hatchet," Rose says. Still, he admits it gave him some strange satisfaction when his favored procedure was chosen over von Braun's.

Rose would openly argue with close acquaintances as well. Owen Maynard, who had migrated to NASA from Canada about the same time as Rose did, was a marvelous engineer. He helped develop the initial designs of the command module and in 1963 was leading the development of the lunar module. As Apollo was being hatched, Maynard came up with a mission schedule he labeled A through G, with A being an unmanned test launch and G, a lunar landing. Most of the plan made sense to Rose, all except for Mission E, initially described as a "deep space evaluation." Maynard wanted to fly a spacecraft *almost* to the moon, but he did not want the craft to circle it on a free return trajectory or enter lunar orbit.

Rose told him the idea was foolish.

"I really took after Owen about this crap of going out near the moon and coming back on a big loop," Rose says. "I told him, 'It's a waste of time, and you're putting lives at risk for no reason.' He said, 'But we want to know how everything reacts to deep space.' I argued [we should] at least do a free return, but he really pushed for the E mission."

Rose had a plan of his own: Forget this deep-space-evaluation business. Why not send the command module and lunar module to the moon into lunar orbit, fly the lunar module down to about 50,000 feet, and check out the spacecraft's maneuverability and rendezvous capabilities in the environment in which it was meant to fly?

"We had never taken the lunar module to the moon," he says. "We had never performed rendezvous around the moon. I said, 'We're not sure whether we've got all of our rendezvous procedures and trajectories right. We want to make sure we've got it all—timing, communicating, the whole nine yards.'"

He went to Kraft with the idea. Kraft liked it. Rose wrote a memo, and Kraft signed it, saying they *insisted* on this mission. The deep-space-evaluation mission became Apollo 10. The flight, Rose contends, has never received the credit it deserves.

Rose admits it was difficult telling three ambitious pilots—all veterans of the Gemini program—they would fly within nine miles of the moon, but not land. But that was his job and one of the reasons he made it a personal policy not to get chummy with the astronauts. "I had been part of airplane testing, and I knew sooner or later we might lose some of them," he says. "I couldn't let my emotions get in the way of sound judgment. I didn't get too involved with the emotion of who was going to be the first person to land on the moon. I was like, 'Whoever it is, it is.' I was more concerned with the operations aspect of it. I had to detach myself from it and go from there. I certainly got emotional later on, as 11 was landing and all that, but not as we were doing the planning."

Still, he felt for Tom Stafford, the commander of Apollo 10. "Bless his heart, Tom had his heart set on landing," he says. "But he said, 'I understand what you're trying to do. I'm for it. Let's do it.' He didn't have to do that, you know. I've always thought highly of Tom Stafford for that. What a temptation it must've been to fly within 50,000 feet of the surface. You could've spat on the damn thing from there. But I told Tom, 'You're doing a yeoman's service here.' He said, 'I know. Won't many people realize it, but I do.' I said, 'I do, too.'"

Rose liked the combination of Stafford, Gene Cernan, and John Young. He says Cernan, the lunar module pilot, "was a born leader" and calls command module pilot John Young, "the epitome of what a test pilot should be . . . cool and doesn't get flustered."

In May 1969, two months before Apollo 11 was scheduled to attempt the first landing, Stafford and his crew spent more than two days in lunar orbit. Stafford and Cernan maneuvered their lunar module, *Snoopy*, to within 50,000 feet of the surface, then successfully performed the critical rendezvous procedures to reunite them with Young and the command module, *Charlie Brown*. It was exactly what Rose had hoped for—a test drive that provided valuable information for Apollo 11 and the first moon landing.

"One of the things Apollo 10 did, apart from getting all the procedures sorted out, was discover that there were some lunar mass concentrations [mascons]—gravity effects—that the unmanned vehicles didn't detect," he explains. "They perturbed the trajectory, and you have to realize that when you're flying down around the moon, you only need a little perturbation to put you off quite a bit. We had to allow for the effect [the gravity effects] would have on the spacecraft and configure that into the guidance system.

"I hate to think what would've happened on Apollo 11 if we didn't know about those mascons. There were enough problems as it was, with Neil having to take over manually to fly over boulders and craters that we didn't know were there before. We didn't have the high-resolution photography that we have today, so we hadn't seen them. He had to do so much manual flying, he only had a few seconds of fuel left. Without being able to allow for the mascons, there's no telling where Neil and Buzz would've wound up."

Fate was not kind to Tom Stafford. He never flew to the moon again, but both his crewmates would later command missions and walk on the lunar surface.

In 1964, Rodney Rose's knowledge of radiation was limited to the x-rays he had received over the years while dealing with nineteen kidney stones.

"But one day Chris Kraft stuck his head in my office door and said there was a big argument going on between the sun physicists and the medical people about radiation. He asked me to go take a look at it."

It was a typical NASA argument for those days—radiation experts on one side of the table, a team of physicians on the other. The first thing Rose did was to organize a Radiation Constraints Panel to study the problem, which was real. Astronauts headed to the moon would pass through the Van Allen belts. These fields of radiation encircled the earth at varying levels. Rose felt comfortable that the astronauts would be traveling at such incredible speed—approximately 36,000 miles per hour—their exposure would be minimal. "We zipped through them in a matter of minutes," Rose says.

The major concern was solar flares—massive explosions on the sun that create huge amounts of radiation that drift out into space—and what to do, should one or more occur during a lunar mission.

Explains Rose, "If an event occurred on the west side, it wasn't a big deal. But if it occurs on the east limb, the particles are going to be placed in the earth-moon area of space. That was something we had to address. So we set up procedures for the lunar landings and for Apollo 8 and 10. If there was a major event when they were at the moon, they would stay there until [the radioactive particle] had decayed a bit. And they would turn the command module so that the hatch and the windows were facing the moon."

The moon would protect the spacecraft for half of each orbit; during the other half, when the spacecraft was exposed, the service module and engine would serve as a shield.

"If they were on the lunar surface, and we had data capability of having twenty-four-hour advance notice before the major particle events reached the moon's surface, the astronauts would immediately leave the moon and rendezvous with the command module," Rose says. "And they would stay there until the worst of the particles had passed. They might get some [radiation], but not enough to make them sick or vomit or anything."

Rose pauses. "You know, somebody must've been with us on Apollo 11. I've got a plot of that flight, and we had about five or six major events occur during July 1969, but not a single one while we were flying. They happened either just before the flight or right after the crew returned. We were just lucky. That's sort of the way Apollo went."

Rose, who is retired and lives near Houston with his wife Leila, is still asked, "How did we do it? How did we land a man on the moon just eight years after man's first journey into space?"

"It was political," says Rose, the father of two sons. "[President] Kennedy realized he had to unite the country to do something that would beat the Russians, because they had been up [in space], and we had had some marvelous failures with our unmanned launches. Three or four of them just cratered, right there on the launch pad. So we *had* to improve and we *had* to excel.

"You know, the very fact that we landed on the moon our very first go at it was really incredible. Thing is, in those days we had built one heck of a good team of contractors, government people, and so on. Everybody was dedicated to the job, which you don't see much now. We had one blinding ambition—get to the moon, land there and come back safely. And when we did it and beat the damn Russians, it was almost like winning a war, at the time. It's hard for people to understand today, but that's the way it felt at the time."

Rose remained with NASA twelve years after Apollo ended in 1972, working the shuttle program as a mission planner and in data systems and analysis. At his retirement party, he was presented an American flag that was aboard Apollo 17, the program's final journey to the moon.

"I'm just thankful to have been a part of the whole moon thing," Rose says. "I mean, there will never be another Apollo 8—going to the moon for the first time, orbiting it on Christmas Eve? I've heard Hollywood is considering a movie about it. I say it's about time. It's got all the drama anyone could ever want.

"The sad thing is, Apollo shouldn't have ended. It was disgraceful the way they did it. We had placed three Apollo Lunar Surface Experiment Packages [ALSEP] on the moon [on Apollo 15, 16 and 17] with a 126-year lifetime. They were sending us data about everything occurring on the moon. It could've told us a *jillion* things about the moon. And it cost only about $5 million a year to keep up a small controller to record it. But Congress just stopped the missions, then stopped the funding for the ALSEPs. Once we cut them off, there was no way to turn them on again. So they're just sitting up there. It is criminal. A tragedy. And now they're talking about going back to the moon? Just think how much we would know if those ALSEPs were still up and running."

Rose often encounters people who believe exploring space is a waste of money on a planet that has millions of people starving. He says such narrow thinking is what will finally destroy the earth.

"I try to tell people that eventually we're going to have to evacuate the earth," says Rose. "Sooner or later an asteroid is going to hit the earth. There are roughly 500 asteroids that intersect the earth's orbit [around the sun] every year. Now, the fact that one hasn't hit us yet is because the area between the earth and the sun is a damn big place. But we're playing Russian roulette. We've got to come up with a plan to be able to move human beings to another place—the moon or wherever. So we'd better find a place to do that.

"I have a hard time getting people to think along those lines. I guess I've always been a bit of a 'far-out' planner on stuff like that."

13 GERRY GRIFFIN

Flight Director

Gerry Griffin spent four years in Sunnyvale, California, in the early 1960s, helping launch satellites into orbit for the military. The Satellite Test Center was an exciting place to be, especially those precious moments when a rocket lit and thundered successfully into space. He enjoyed the team atmosphere and the mechanics of the control room, and he became recognized as an expert on the rocket's upper stage, known as the Agena.

But Griffin's heart was elsewhere. He desperately wanted to work for NASA.

"With the lunar missions being the ultimate goal and America in a race with Russia to see who could get to the moon first, I knew it would be a real hoot, if we could pull it off," he says.

Griffin, who was married with two children, had interviewed with NASA several times; they could never agree on salary or a position. But as the space program progressed quickly from the one-man Mercury missions to the two-man Gemini flights, Griffin decided he'd better latch on wherever he could, if he wanted to be part of the moon journeys. So in 1964, NASA hired him—and sent him back to the Satellite Test Center to continue working on the Agena, which had just been chosen as the rendezvous target vehicle to be used during the Gemini program.

"I had taken a pay cut to sign on with NASA," Griffin says, "and here I was right back where I had been. But at least I had my foot in the door. And it was the best step I ever made."

He had barely settled into his old chair in Sunnyvale when NASA called with news so wonderful, Griffin worried that it was a prank. "They

Flight director Gerry Griffin (*right*) discusses a procedure with Christopher Kraft, director of flight operations at NASA's Manned Spaceflight Center. [Courtesy Gerry Griffin]

said they were going to move me over to Gemini flight operations because they needed some people who had been in a control center environment," he says. "Well, I'd done that for four years, so I took to it like duck to water."

He moved to Houston, where the Manned Spacecraft Center was under construction, and got to know the top guns of flight operations—Christopher Kraft and Gene Kranz. He quickly learned that, while it naturally had a sharper edge because human lives were involved, manned spaceflight was much like what he had been doing at Sunnyvale. "It was all nuts and bolts and volts and ohms and amps—getting the spacecraft right," Griffin says. "And we were all learning. What we were trying to do had never been done, so there weren't a whole bunch of rules laid out for us. We were sort of writing our own script, and you don't get to do that very often. It was fun."

But one afternoon, while sitting inside a makeshift control room, Griffin glanced around at his workmates, who were all wearing headsets and plotting paths no humans had ever traveled. Something suddenly occurred to him. "I might be in this room—or one like it—with these people when we go to the moon one day."

He was.

Griffin would go on to serve as a flight director—the "head coach" of mission control—on ten of the eleven Apollo missions. (He was part of the mission control support crew on Apollo 8.) And he was one of the few who didn't speak out angrily when Congress pulled the plug on the program in December 1972, following Apollo 17. It had been a grueling eight years, he admits, and no immediate plans were on the table to build a moon base or go to Mars.

"As it was, I'm not so sure what else we could've proven," says Griffin, who now serves as a technical and management consultant to several corporations. "We'd won the race with the Russians; so that wasn't an issue. We could've gathered more rocks, taken more pictures. Maybe if we'd had a whole lot more time, we could've done a landing in the huge impact crater, Tyco, and probably gotten some stuff that would have been very revealing—Jack Schmitt [Apollo 17 astronaut/scientist] was pushing hard for a Tyco landing. Or we could've done a landing on the backside. But both of those would've taken a lot more money, different systems. . . . We would've had to put a satellite up just to be able to communicate with the astronauts, if they'd landed on the backside. No, I think it ended about the right time. Another landing or two wouldn't have added much to the database.

"We landed six times and never left a man up there. Sometimes I think we made it look so easy that people started taking it for granted. I didn't. In time, others won't either. What we did was amazing."

Griffin and his twin brother Larry were born on Christmas Day of 1934 in Athens, Texas. Their dad, Herschel, worked as an abstractor, recording the history of land transactions. "His family had farmed growing up, so I think he'd had enough of that by adulthood," Griffin says.

Their mother, Helen, taught school. She died at the age of forty-two after a two-year battle with cancer. "I remember it was all a little bewildering," Griffin says. "And when she died, I don't think I was very surprised. I remember that it *really* brought Larry and me closer to [brother] Ken, who was ten years older. In fact, Larry and I shaped our lives after him."

Ken had entered the U.S. Army Air Corps when he was nineteen and flew B-17 bombers during World War II. Later, he became the first member of the family to graduate college.

"To us, Ken was bigger than life," Griffin says. "So we knew we wanted to fly planes and get an education, just like him."

Gerry and Larry earned aeronautical engineering degrees in 1956 from Texas A&M. Both then joined the Air Force.

During his physical exam, Gerry was diagnosed with 20/25 vision—just poor enough to keep the military from trusting him with one of its planes. "But I was getting into a jet somehow," he says, "so I became a weapons-system officer in the backseat of fighter interceptors, like the F-101 Voodoo."

The time he spent in those jets proved invaluable to Griffin once he arrived at NASA. By that time, he also had earned his private and commercial pilot's license.

"Pilots have a certain lingo. They just communicate differently," he says. "Kranz was a pilot during Korea, and he's mentioned many times that he had an affinity for the people who had spent time flying. A lot of the flight directors were pilots, and I don't think that was a coincidence."

When it came to picking flight controllers, Kraft and Kranz knew what they were looking for. "They wanted people who were comfortable in an environment where you had to make rapid decisions, and you had to be right 99.8 percent of the time," Griffin says. "It was a lot of pressure, but most of the guys thrived on it. We had a few who didn't make it, but they kind of took themselves out of it. They realized they didn't have the stomach for it.

"If you were going to survive, you had to be the kind of guy who didn't mind standing out there on the end of the diving board without anyone else around, then be cocky enough to say, 'I'll make the right call. I'll do it right.'"

Griffin credits Kraft for a lot of the mental toughness that developed inside mission control. Kraft, a Virginian with an engineering background, had been a member of the Space Task Group, formed to speed up America's efforts after the Soviets became the first in space in 1957, with their unmanned satellite, Sputnik. Kraft's sense of urgency rubbed off.

"He reminded me of Bear Bryant, who came to A&M as football coach my last season there," Griffin says. "I know a lot of the players who played for him, and they all say the same thing. They weren't scared of him. Well, maybe a little. But they said one of the reasons they wanted to play well was . . . him. They didn't need a lot of adulation from the press. All they needed was a pat on the back from Bryant once in a while.

"And that's really the way we were about Kraft. He was tough. Mostly fair. He was always fair with me, but I saw him get a little too harsh with some other people, I thought. But he just had this magnetism about him, like Bryant. If he ever came down and put his hand on your shoulder and said, 'That was a good job, young man,' you could take the *New York Times* and the *Houston Chronicle*, and every other newspaper, and forget what they said about you. If Kraft thought you did something right, it was like you were walking on water."

Griffin spent his Gemini-program time on mission control's front lines as a guidance and navigation officer. When Apollo was ready to go in October 1968, with an eleven-day, earth-orbit marathon to check the command module's systems, Griffin had been promoted to flight director. He was thirty-four years old and supervised a team of up to 100 specialists.

"Looking back, that was quite a position for a guy that young," he says. "I don't mean that in a boastful way. But I could move a recovery aircraft carrier from one ocean to another. Move airplanes around the globe. You had to be focused, or it could go to your head."

They had simulated it dozens of times, if not hundreds. But no one in mission control was fully prepared for the emotion that filled the room when capcom Michael Collins radioed this message to the crew of Apollo 8: "You are GO for TLI."

Trans-lunar Injection. Frank Borman, Jim Lovell, and William Anders were about to leave earth orbit for the moon. They would spend twenty hours in lunar orbit from about sixty miles high, the bulk of it on Christmas Eve, 1968.

"You could've heard a pin drop in that control center," Griffin recalls. "And when the burn was performed—and it cut off right on the money—we all kind of just looked at one another. Without uttering a word, we all said the same thing. 'We just did it. We're headed to the moon.' This was not a situation where you could say, 'OK, we're gonna stop now.' We had put them on a trajectory where they had to go *out there*."

On the way to the moon, Borman began vomiting. This was not unusual. Many astronauts who flew in space dealt with some degree of nausea. But everything was magnified because, for the first time in history, humans were outside the earth's influence.

"I think we all started wondering, 'Maybe there's something about going to the moon that we don't understand,'" Griffin says. "It was an unknown that we had no answer for."

Borman's queasy stomach, which eased before the spacecraft reached the moon, turned out to be about the only glitch of the mission. The crew beamed back TV pictures of the lonely lunar landscape. And though he didn't dare show it, Griffin got a real kick out of hearing Anders' voice from up there. They had flown together in the Air Force in the late 1950s. "And now this guy was 250,000 miles from earth," Griffin says. "That was neat."

Griffin felt the same kind of awe the night Neil Armstrong and Buzz Aldrin took the first steps on the moon. "My team was working the sleep shifts on Apollo 11," he says. "So after they got out on the surface and did a few things, I had to go get some sleep because I had to be back pretty soon. I walked outside, and on that clear July night in Texas, there was the moon sitting right above the control center. I was like, 'Holy shit! Those guys are up there, and we've got to get them back!'"

But there was also a sense of victory. America, which had been woefully behind during the early days of the space race, had beaten the Soviets to the moon. This fact was not lost on one of Griffin's team members who wrote, "THE RUSSIANS SUCK" and projected the words up on one of the screens in the front of mission control.

"We all got a good laugh out of that," Griffin remembers, "but I finally said, 'Guys, we probably need to get that down before the press sees it.'"

As a flight director, Griffin had tried to remember the number one thing Kraft emphasized: Pick good men for your team; then let them do their jobs. On Apollo 12, his first as head flight controller, he didn't have to wait long to test how much he believed Kraft's theory. Thirty-six seconds after liftoff, then again twelve seconds later, the Saturn V rocket and the command module carrying Pete Conrad, Dick Gordon, and Alan Bean were struck by lightning. All the astronauts and flight controllers knew at the time was that something had made the electrical system go berserk. To the crew, it seemed like every warning light on *Yankee Clipper*'s instrument panel was glowing. The guidance platform was lost, but Apollo 12 thundered on as if everything was normal.

His head told him, "Abort—now!" But Griffin, for reasons he still doesn't fully understand, stood quietly for a few seconds and then radioed twenty-four-year-old John Aaron, head of the electrical system. "What do you see?" Silence. "What do you see?" Griffin repeated. The truth was Aaron saw nothing. His console was blank.

"I had been a GNC [Guidance Navigation Controller] before I was made flight director," Griffin says. "So I had a total understanding of those systems." But even he was surprised when Aaron said to capcom Jerry Carr, "Flight, try S-C-E to Aux." (Aaron had recalled a similar problem during a simulated launch months earlier.) Bean flipped the switch, and Aaron's console filled with data. He was back on-line.

Once the second stage fired, Bean reset the fuel cells—the command module's main source of power that appeared to be damaged—and the warning lights went dark. The fuel cells were up and running again.

One major problem remained as the astronauts reached earth orbit: "The IMU [Inertia Measurement Unit] had tumbled and lost its way," Griffin says. "So the eight ball—the attitude indicator in the command module—was just going around in a circle." Without it, the spacecraft and the astronauts would have no sense of direction. Everyone involved understood the situation: Fix it or scrub the mission.

"I remembered that you could damage the IMU—the gimbals in there that this thing floats around on," Griffin says. "If it tumbled, you were to pull a circuit breaker, and it would lock it. And then you could power it up and get realigned later."

On the voice tapes, Griffin can be heard saying to Aaron at least four times, "Don't you want to pull the circuit breaker?" But each time, Aaron responded, "Stand by, Flight."

"He was running it through his backroom guys, and he had the contractors on the line," Griffin says.

Aaron was working the problem, just as he had been drilled to do. Only a few minutes passed before Aaron radioed to Griffin. "Flight, have him pull the circuit breaker." Griffin relayed that to Carr, who then radioed the command to Bean.

"It worked," Griffin says. "I was afraid we might have waited too long and had damaged [the IMU]. But we got it back aligned, and it worked fine the rest of the mission. We handled the problem the way Kraft had taught us to. Those guys, like John Aaron, were in those seats for a reason. And as long as we communicated properly, the system Kraft had developed would work."

Apollo 12 went on to a pinpoint landing in the Ocean of Storms, touching down within 600 feet of the unmanned probe Surveyor 3, which the United States had landed there two years earlier.

"In fact, the only other thing to go wrong was when Bean accidentally turned the TV camera lens toward the sun and knocked out live television for most of the EVAs," Griffin says.

Being a loyal Texas A&M Aggie, Griffin still reminds the University-of-Texas graduate about his error, every chance he gets.

Apollo 13, a blockbuster movie in 1995, remains one of actor Tom Hanks' biggest hits at the box office. It has grossed nearly $200 million—more than *The Green Mile, Sleepless in Seattle, Forrest Gump*, or *A League of Their Own. Apollo 13* tells the story of the flight that nearly left three astronauts stranded in space, after an explosion on the way to the moon.

"From a technical standpoint, the movie was right on the money," says Griffin. (Griffin served as a consultant to Hanks and director Ron Howard during the filming; he later worked as an advisor/actor in *Contact* with Jodie Foster and *Deep Impact* with Robert Duvall and Morgan Freeman.) "There are some things they had to do with dramatic license. Like at the start of the movie, there's a big party on the event of the Apollo 11 landing. That party never happened, but it was a way of introducing all the characters at the front of the movie. Then they could get on with it."

Another dramatic license led viewers to believe that Gene Kranz was the only flight director on duty for the entire six-day mission. Not true, of course. Griffin, Glynn Lunney, and Milton Windler led teams that shared rotating shifts with Kranz's, but they were nowhere in the film.

"I knew what was coming," Griffin says. "I had talked with Ron Howard when he and Tom Hanks were trying to decide if they wanted to make the movie. Ron knew a lot about the flight. So did Hanks. They had studied a lot about it. But Ron told me they simply didn't have time to develop four characters as flight directors. And he said since Kranz was on duty when the explosion occurred, he was going to make a 'composite' flight director [of] him. He asked if it bothered me, and I told him, 'No, as long as you keep the technical parts correct.' He said, 'That's why I've got you,' and he asked me to be a consultant.

"I understood his situation. If they'd wanted to make me a star, that would've been fine. But I wanted to make sure the guts of the movie were right and it wasn't some far-fetched treatment of what really happened."

Actually, Griffin would have made an interesting character, the night of the explosion: "I had just gotten off shift, and my team turned it over to Gene's team. I got out of my chair, and Gene took it." Then Griffin went and played third base for one of NASA's recreational softball teams. He had stopped by NASA engineer Ed Fendell's apartment after the game for a beer when Griffin's wife, Sandy, called.

"She said, 'They're trying to reach you. They've had a problem,' I asked what had happened, but she wasn't sure," says Griffin. "So Fendell and I called in and were told it was some sort of oxygen problem. I went right over to the space center still dressed in my softball uniform."

As soon as Griffin walked in, he could tell the problem was serious, "just by the look on people's faces." About that time Lovell radioed mission control that he could see the spacecraft venting something out into space. It was the oxygen. Jim Lovell, Fred Haise, and Jack

GERRY GRIFFIN

A flight director during Apollo, Gerry Griffin served as a technical consultant on the movie *Apollo 13*. [Courtesy Gerry Griffin]

Swigert were 200,000 miles from earth, and their command module, *Odyssey*, was all but dead. They wouldn't know it until after they returned to earth, but an oxygen tank in the service module had exploded, causing a chain reaction of failures. *Odyssey* was quickly losing its electrical power—and virtually everything in the command module ran on electricity.

In perhaps the most tense moments of any Apollo flight, the astronauts powered up the lunar module, transferred the guidance platform over to its computer, and took refuge in the thin-skinned, spider-like craft. They shut down *Odyssey* to save what power was left for reentry and to give ground controllers time to try and come up with a new game plan. All of this was virgin territory. The lunar module wasn't designed for three astronauts or to serve as a primary vehicle; it was for landing two astronauts on the moon, then returning them to the command module. Nothing more. And no one knew if a command module could be shut down in the icy vacuum of space and brought back to life. One thing was for sure: They wouldn't be landing on the moon. Getting the astronauts back safely was now the sole mission.

"When Ron Howard and Tom Hanks were talking to us, when the movie was still in the planning stages, they asked us, 'Weren't you scared?'" Griffin says. "It was about a half dozen of us in there—me, Kraft, Kranz, Lunney. I don't remember who else. We kind of looked at one another and said, 'Naw, scared is not the right word.' And it wasn't because we had been trained that as long as you've got options, [you can] just keep using them until you use them up. And we never ran out of options. We got close, but we never ran out. There was a workmanlike approach of 'Here's what we've got; what do we do?'

"I know I didn't sleep at all. It's not in the movie—well, it is, but you really can't tell it—but we actually took Kranz and his team off the consoles and sent them back to that [meeting] room, shown in the movie, to try and figure out what to do. They were on when they powered down everything in the command module and powered up the lunar module. They were the best to figure out what position all the systems and switches were left in. So for the next three days, the other three teams actually manned the control center."

Kranz's outfit, along with contractors and engineers who had helped design and build the command module and lunar module, worked on procedures to get the astronauts home as soon as possible. The other flight teams manned the consoles and kept a close watch on consumables—electricity, water, oxygen, and food.

"At first, we thought oxygen was going to be our shortest supply," Griffin says, "but it wasn't. It was water. When we lost the fuel cells [in the explosion] we couldn't make any more water."

Up in *Aquarius,* Haise had come to the same conclusion. He knew more about the spacecraft than most in the program. Haise had spent nine months, out of one fourteen-month period, working 24/7 on factory tests of all the early lunar modules. His knowledge far exceeded what was needed to fly the vehicle, and it provided him with great confidence that *Aquarius* could meet the extreme demands suddenly thrust upon it. But the problem wasn't *Aquarius*; it was the amount of water available. By Haise's calculations, they would run out five hours *before* reentry. Water was critical, not just for drinking purposes, but because it cooled the lunar module's systems. Somehow, the crew had to make up those five hours, hopefully with room to spare.

As *Aquarius* headed for a slingshot path around the moon and back toward earth, ground controllers wrote a procedure for a five-minute burn that would be performed two hours after rounding the moon. It would cut the crew's trip home by about eight hours. But the burn was moot if *Aquarius* wasn't aligned properly, and only the astronauts could verify that.

Griffin listened anxiously from the head chair in mission control while Lovell and Haise looked through the Alignment Optical Telescope as Apollo 13 neared the moon's shadow. "If they could see the sun at all [through the eyepiece], then we knew we were good," Griffin says. "If not, we had a whole new set of huge problems."

When Griffin heard Haise say, "Yes . . . upper right corner of the sun," he wanted to shout for joy. As he wrote "AOT ok" in his logbook, his hands shook so badly, the letters looked like a preschooler's scribble.

Some sixty-seven hours later, Apollo 13 floated to a soft landing in the Pacific. Griffin describes his emotions as "All right! Yes!"

"I think we all still felt bulletproof at that point," he adds. "But I've thought many a time about the timing of the explosion. What if it had happened after Jim and Fred had landed [on the moon]? Then we wouldn't have had the lunar module or its descent engine. Timing was so critical."

After Apollo 14 successfully landed and explored the moon for more than nine hours, Griffin was back in the lead flight director's seat for Apollo 15. And he was armed with a different attitude.

"Up until then, the people in the control center, and even the crews to a certain extent, had been focused on getting up there and getting back," he says. "That had been our primary purpose. But when we got to 15 . . . the thinking had to change a little bit. It was time to *really* explore. [Apollo 15 commander] Dave Scott sort of sensed it, and he brought it to my attention. I knew he was right."

Apollo 15 would be the first mission to use the Lunar Rover, a "space car" that would allow a much wider range of exploration. It would also be the first mission with three EVAs, totaling more than nineteen hours.

"Dave called me one day and said, 'If you're going to be doing all our EVA stuff, would you like to go out and train with us, see what we're going through, and how we do our stuff?' So I went with them out to New Mexico and California," Griffin says. "And it really opened my eyes to the mechanics of what it took for them to get those samples and document them correctly, with the right kind of lighting, so they could photograph them before they put them into the bags. And I listened to the geologists telling them, 'You really ought to be looking for this kind of rock and that kind of soil.' And it dawned on me that there was now a completely different set of issues for Apollo than I had dealt with. And it really turned me on.

"No doubt, it helped me do a better job on the final three missions. It just gave me a better feel for what we were gonna try to do during these EVAs."

Griffin laughs. "I remember on some of the earlier flights when some of the science guys wanted [the astronauts] to go and do something, I'd roll my eyes and think, 'A rock's a rock. Go get something and get back in!'"

During the final EVA of the program on Apollo 17, Griffin had the difficult task of cutting it short. Eugene Cernan and Jack Schmitt had explored the moon for twenty-two hours, covered nearly nineteen miles, and collected two hundred and forty-three pounds of samples. They were exhausted. Tools had rubbed their hands raw. But the geologists begged for just a little more time.

Griffin refused. Cernan and Schmitt had done their jobs well. So had everyone in Apollo. He wouldn't take the chance of something going wrong because an astronaut overexerted himself trying to get *one* more sample. He ended the EVA.

"I was happy," Griffin says, "and Schmitt was happy. And that was a key sign for me. When your scientist-astronaut is happy, you know things have gone well."

After passing the rock boxes up to Schmitt in the lunar module, Cernan finished what Neil Armstrong had started three years earlier.

Gerry Griffin was the lead flight director on Apollo 17, the program's final journey to the moon, in December 1972. [Courtesy Gerry Griffin]

"We leave as we came," said Cernan, glancing at the earth hovering in the southwestern sky. "And, God-willing, as we shall return, with peace and hope for all mankind. Godspeed the crew of Apollo 17."

Many Apollo astronauts and top ground people gathered in the summer of 1975 to ponder their accomplishments. "Jack Schmitt had a fellowship [at Cal Tech] that had some money left in it. It paid to get a few of us out there," says Griffin, now a grandfather of three. "You have to understand, there had never been any of that touchy-feely stuff about Apollo. But we finally asked ourselves, 'How did we do it? How did we pull it off? And why?' Kraft answered that last one. He said, 'We didn't go to the moon because we were trying to get to the moon. We went to the moon because where we *really* want to go is Alpha Centauri'—the closest star, which is light years away. He said this was just a little baby step.

"We talked about the space race and agreed that we probably would've wound up doing something like this, had there not been a race, anyway. But the race with the Soviets just made it happen faster."

Every person took a turn, said what was on his mind. Then Neil Armstrong slowly got out of his seat and walked to a blackboard in the front of the room. Griffin had no idea what he was about to say, but he knew it would be good.

"I think all of us were happy to hear that Neil was going to be the first guy on the moon," Griffin recalls. "All the guys around me were like, 'Hey, that's a *good* selection.' We all had great respect for him. He wasn't real talkative. But I used to say this about him: When he says something, you'd better listen, because it's generally important and . . . accurate.

"In his X-15 days, he was known as a great test pilot, very thorough. And he proved his stuff on the Apollo 11 landing, going down into that boulder field, taking over manually, and scootin' over it until he could find a smooth place to touch down. And he did it in such cool fashion. You didn't hear him talking much about it. That's just the way he handled things, with a cool head."

As Armstrong stood at the blackboard that afternoon in California, the room fell silent. He picked up some chalk and drew four curves that looked like mountains.

"He labeled each one," Griffin says. "He labeled one A THREAT, which the Soviets were. He labeled the next one BOLD LEADERSHIP, which we had with President Kennedy when the space program first got going. Next he wrote WORLD PEACE, and the Vietnam War really hadn't started when NASA first began in 1958. Lastly, he wrote

TECHNICAL SAVVY. Then he put down the chalk and explained what they all meant.

"He said, 'When you get these curves lined up, all happening at the same time, something such as Apollo is going to occur. There's nothing you can do to stop it. And that's why we did it. Because we had all those factors lined up for us at the same time.' We all started nodding our heads because he was right. And I use that a lot, particularly when I speak to young groups. I tell them what Neil said, that they have to be ready because one of these days, those curves are going to line up again.

"But we had all that and more. We had the American people really pulling hard for us, even after losing three astronauts in the Apollo I fire, when some countries might have backed off. The support never wavered—not in Congress, not among the people. And that meant more than they realized."

14 SAVERIO "SONNY" MOREA

Project Director, Lunar Roving Vehicle

NASA seemed to think of everything while planning its uncharted path to the moon, from developing a rocket powerful enough to propel humans out of the gravitational forces of earth, to figuring out what to do with nine days worth of three astronauts' body waste. The engineers and scientists imagined the needs, technical and practical, well in advance, and developed solutions.

Except for at least one: If astronauts were going to make the most of their time on the moon and explore as much of the lunar terrain as possible, they needed a car. "Unfortunately, no one thought about it in 1960, when they could have developed it in leisure," says Saverio "Sonny" Morea. "They thought about it three months before the first landing."

So in May 1969, Morea was placed in charge of developing a moon car. Bids were taken. When Boeing was awarded the project in October—three months after Apollo 11's historic touchdown on the moon—Morea's team had seventeen months to test and deliver a Lunar Roving Vehicle (LRV) to the Kennedy Space Center in time for Apollo 15. It was a ridiculous demand; most projects of that magnitude took four years to develop. But Morea didn't have time to whine. He recalled a saying that began traveling around NASA shortly after President Kennedy issued the challenge, in 1962, to reach the moon by the end of the decade. "One day of slippage in any project—just *one day*—created a one day slippage in our ultimate goal," he recalls. "Every day was precious."

Morea refused to be intimidated by the timetable or entertain any thoughts that NASA had given him an impossible assignment. He didn't

Jack Schmitt, Ann Morea, and Sonny Morea share a laugh during the thirtieth anniversary celebration of Apollo 17. Schmitt, lunar module pilot on the mission, used the car that Morea was in charge of designing to explore the moon for miles away from the lunar module. [Courtesy Sonny Morea]

want to hear excuses from his team or anyone else—not even, upon his return from Apollo 11, Neil Armstrong.

"There was some doubt as to whether we could even drive on the moon," he says. "When Neil got back, I sat in on the debriefings, and I asked him, 'What kind of vehicle do we have to have to get around up there?' He looked at me and said, 'There are craters everywhere. The tires would have to be twenty feet in diameter!' Honestly, I think he was so overwhelmed by his moon experience that he was exaggerating and didn't even realize it. He was trying to make the point: 'There's a lot of craters up there, fella. The bigger you make the wheels, the better you're going to be.'

"It would've been nice to put a car up there with wheels twenty feet in diameter, but not possible, of course. So I quit asking [Armstrong] questions, because his answers weren't doing me any good. I had a job to do, and I didn't need somebody telling me it couldn't be done."

Morea, born and raised in New York City, came from what he terms "very humble beginnings." His father was a lifelong construction worker. His mother made dresses for extra income.

In high school, Morea got a job with his father every other weekend, earning $13 a day pushing a wheelbarrow. "On the in-between weekends,

I'd take that money out to a small airport and take flying lessons," he says. "By the time I graduated, I had my pilot's license."

He was an exceptional student. "Had my parents had the money, I would've gone to New York University," he says. Instead, he graduated in 1954 with an engineering degree from City College of New York and was also commissioned as a second lieutenant through the school's Army ROTC program.

His first job was with North American Aviation in Los Angeles—a company now known as Rockwell International. Morea worked as a designer on the Navaho Project, one of the early rocket programs that laid the groundwork for future success in the space program.

After a year, Morea was called to active duty and trained several weeks at the Aberdeen Proving Grounds, a military weapons research center in Maryland. Then he received orders to report to Redstone Arsenal in Huntsville, Alabama.

"I didn't even know where that was when I got the letter," Morea says, laughing. "I was confused. I had no idea what was going on down there and didn't know why they had picked me."

Morea soon learned he was headed to the nation's top rocket research facility. Upon arrival, Morea was summoned to a meeting with the facility's director, Dr. Wernher von Braun.

"On my way down to Huntsville, I heard that there was this crazy German scientist who was now helping us develop our military vehicles," he says. "So I had a vague knowledge of who he was.

"When I walked into his office, I was just overwhelmed by the power of the man. He was a physically striking man, a big man. Intelligence oozed from him. But on top of that, he had a wonderful personality. He was kind, humble, very down to earth."

Von Braun had led the development of the powerful V-2 German missiles that pounded London during World War II. He was proving to be one of Hitler's strongest assets, but von Braun wanted no part of the war. He was arrested for concentrating his efforts on rockets that could orbit the earth—perhaps even fly to the moon—rather than on helping bolster Hitler's arsenal. Von Braun eventually was released from jail with the idea that he would work on Germany's war efforts. Instead, he surrendered to U.S. forces in 1945, bringing with him a team of scientists and engineers, plus his own rocketry expertise, unmatched by anyone in the world.

By 1950, von Braun headed the Army's development of the Redstone rocket, which launched America's first space satellite, Explorer I, into earth orbit in 1958 and also powered Alan Shepard and Gus Grissom on their fifteen-minute suborbital spaceflights in 1961. Von Braun was living his dream in Huntsville, and Americans were reaping the benefits.

"To tell you the kind of person Dr. von Braun was," Morea says, "much later in my career, I had to be down in Birmingham with my family on a Sunday morning. I had flown into the airport, and we were about to go grab some lunch. I was waiting on my ride, and as I [looked] out the window there [was] Dr. von Braun with his wife [Maria] and son [Peter]. They walked into the building we were in. It was huge, probably 150 or 200 feet long. And I was all the way at the other end. Well, he had a habit of always looking the entire area over when he walked into a room. And when he looked down at the far end of it, he saw me. He had the perfect opportunity to just walk away; his ride was there. But he called out from the other end. He introduced his wife, his son. That's just a small indication of why people loved working for him. He came out of a totalitarian structure, yet he was as democratic as we would want ourselves to be. There was nothing dictatorial about him."

When Morea was hired at Redstone, von Braun told him to "take three or four weeks" and tour the various areas, then let him know which one he preferred working in. Morea picked propulsion.

But that didn't last long. Ludwick Roth, the No. 3 man on von Braun's team from Germany, who was in charge of all the contractors Redstone Arsenal was using, soon offered him a more high-profile position. "He told me, 'I know you're a young guy from New York, and you have flexibility when it comes to travel,'" Morea remembers. "He said, 'I have a company up in New York that provides the actuating mechanism in the jet stream of the rocket engine, and they're in deep trouble. We need to have a great deal of communication with them. Would you care to be a liaison?' So I did that. And it really helped me in the long run. I would have to give presentations at different places, and Dr. von Braun would be there."

In July 1960, Von Braun was hired as director of Marshall Space Flight Center in Huntsville—NASA's new rocket development center. Four months later, von Braun brought Morea to Marshall as project director of the F-1 engine, which eventually would become the heart and soul of the Saturn V rocket, the key to getting men to the moon.

"The rocket we had at the time, the Redstone, essentially provided about 150,000 pounds of thrust," Morea says. "The Saturn V would create 7.5 *million* pounds of thrust with five of those F-1 engines in the first stage."

In 1966, at von Braun's request, Morea turned his attention to the critical Saturn IVB engine, which would be used in the Trans-lunar

SAVERIO "SONNY" MOREA

Saverio "Sonny" Morea is known for designing the Lunar Roving Vehicle, or the "moon car." But he also helped develop the engines in the powerful Saturn V, which powered astronauts to the moon. [Courtesy Sonny Morea]

Injection burn, taking astronauts out of earth orbit and on a path to the moon.

By 1969, after the Saturn V had performed beautifully on Apollo 8's and Apollo 10's lunar orbital missions, Morea's track record for delivering whatever was asked of him was impeccable. Morea believes it's the reason von Braun gave him the challenge of the LRV.

"It didn't have anything to do with propulsion," Morea says of the moon car project. "But I guess Dr. von Braun figured if you could manage one program, you could manage any of them. It's all dealing with people and resources, and having enough smarts to know when you're in trouble technically, and [being] able to bring the right expertise on board."

Morea could see the need for a car on the moon. But he came to fully understand the LRV's importance in July 1969 as Neil Armstrong and Buzz Aldrin went through their two-hour, thirty-one-minute EVA, which was far less physically demanding than what future missions would require.

"We could tell they were struggling. We could hear them breathing hard," Morea says. "And that meant they were putting forth a lot of effort, which meant they were consuming a lot of oxygen. We found out that night that working on the moon is hard. It was an issue the LRV could help with. If we were going to do any serious exploring, it was a must."

By design, Armstrong and Aldrin covered an area about the size of a standard football field; it was the first landing, and NASA wanted the astronauts to remain within a safe distance of the lunar module. But in later Apollo missions, NASA envisioned exploring chunks of the moon the size of Manhattan. Without a car, that would be impossible.

The first hurdle was figuring out how to include a car on the weight-sensitive lunar module. Every ounce mattered. And a car, no matter how efficiently structured, would weigh several hundred pounds.

Apollo astronauts were not thrilled with the suggested solution—reduce the amount of fuel on board. "It turned into one of the biggest arguments ever in the program," Morea says. "The car wound up weighing 480 earth pounds. So we were talking about taking 480 pounds of fuel away. Well, that 480 pounds could mean the difference in twenty or thirty seconds of hovering time [needed for] trying to find a good landing spot.

"It went back and forth. But at the time, there were some pretty gutsy people who headed the program, namely George Mueller [director of

NASA's Office of Manned Space Flight], who said, 'If you get to the moon, and everything is going well, you're not going to need much time to make up your mind where to touch down. . . . And if you land, what good is it going to do you if you don't have this car?' He's the one who basically paved the way for the car to go along."

Then engineers had to figure out how to build a vehicle that could operate in one-sixth gravity and withstand the extreme lunar temperatures, which ranged between plus or minus 250 degrees.

What sort of tires, for instance, would be functional on the moon? "You couldn't use rubber—it would outgas in those temperatures," Morea says. What sort of steering mechanism would best suit the astronauts who would be wearing bulky gloves? What would be its power source? How would the astronauts remain stabilized in the vehicle as it trekked up mountains and through craters? Would it require an independent navigational source, or could the astronauts simply follow their tracks back to the lunar module?

All these questions, and hundreds more, had to be answered.

"Probably the biggest surprise we ran across during development," Morea says, "was that the astronauts' space suits had such limited mobility. . . . I mean, we had no idea how restricted they were in their movement. Just getting into a vehicle presented problems." When tests were first performed aboard an airplane that simulated one-sixth gravity, it took an astronaut up to ten minutes to get seated in a mockup LRV. Since every second on the lunar surface was precious, this was unacceptable. Engineers made adjustments in the car until the time was reduced to between one and three minutes.

One by one, engineers came up with answers, as they always had in Apollo. The LRV was delivered on time—two weeks early, in fact—on March 10, 1971. Apollo 15 wouldn't fly until late July, but the LRV had to undergo another round of tests before being packed inside one of the lunar module's four equipment bays located on the descent stage.

The finished product—an electric car filled with General Motors parts—looked like a jazzed-up, two-seat dune buggy, with its antennas, communications dish, cameras, toolboxes, and bright orange fenders. "Only reason they were orange," Morea says of the fenders, "is because that just happened to be the color of the material we were using. No special meaning behind it." The vehicle, made primarily of aluminum, was ten feet, two inches long and stood forty-four inches high—about the size of a regular automobile. It consisted of three sections and could be folded into a compact square that measured five feet across and twenty inches thick.

"The safety people stressed the need for redundancy, which complicated the design process, but made it a very safe, efficient vehicle," Morea says.

Two 36-volt, silver-zinc batteries powered the LRV. "The idea was to drive two wheels using Battery A and the other two with Battery B," Morea says. "But one wheel was enough to drive the vehicle, and each wheel could be driven independently. And one battery was sufficient to get the astronauts back to the lunar module in case of an emergency. Again, that was part of the redundancy effort." They would last long enough for the LRV to be driven fifty-seven miles—an odd number arrived at by calculating several variables.

The tires, nine inches wide and thirty-two inches in diameter, were made of piano wire covered with zinc. "They played around with a lot of wire, but piano wires are very lightweight and very sturdy," Morea says. "By putting those Chevron-patterned threads on the outside of the wheel, we got more traction. And the tires wound up only going about a half-inch into the lunar soil."

Instead of a steering wheel, the car was maneuvered with a T-bar controller, located on the small instrument panel—front and center of the astronauts. It provided steering and braking. "The astronauts loved it because it was much like a joy stick in a jet," Morea says, "and that's something they were very familiar with. And since the T-bar could be controlled with one hand, either astronaut could assume driving duties at any time—again, that was part of the redundancy issues."

Speed wasn't a priority on the LRV, although Gene Cernan is credited with the record on Apollo 17, reaching eleven miles per hour down the side of a mountain. "It would've been unsafe if it could've gone faster," Morea says. "Frankly, the astronauts couldn't see very well. When you're on earth and look out toward the horizon, you see various things that give you size relation—cars, trees, people. On the moon, you have nothing, [no way] to judge how large things are. You could be looking at an enormous boulder, and because there's no atmosphere, it would appear a lot closer to you. And when they looked out over the terrain, they couldn't see when they were about to run into a crater.

"It was even further complicated when the sun was directly behind them. You would think that would light everything up in front of you, but there is a point where things get blurred out, and you don't have the depth perception you need. That's why we had to land on the moon in the early morning hours so that the shadow would show you there's a rock over there or there's a big crater over there."

The vehicle's seats looked like beach chairs. In addition to seatbelts, the astronauts were secured with a piece of Velcro on the back of the seats that engaged with strips of Velcro that were on the back of their life support packs.

Safety experts demanded a computerized navigational system be included. "Following the tracks back to the lunar module would've been simple enough," Morea says, "except the astronauts didn't take straight-line routes. And if they needed to get back in a hurry—if one of the astronauts was getting sick, and he was going to puke in his suit and needed to get inside the lunar module and get that suit off—they needed to be able to take a direct path.

"The system wasn't linked to the lunar module. It contained a gyro, much like you find in an airplane. It counted the rotation of the wheels, which helped figure the distance traveled. And it would figure the route—so many feet this way, so many that way. Then it was capable of calculating how many miles it would be back to the lunar module and the proper [straight-line] heading."

Apollo 15 blasted off on July 26, 1971, with Dave Scott, Jim Irwin, and Al Worden aboard. This was the first of the missions to include three EVAs. Morea viewed the launch at Cape Kennedy, then flew to mission control in Houston to monitor the first moonwalk that occurred four days later in the Apennine Mountains.

Scott and Irwin deployed the LRV in approximately thirteen minutes with little difficulty. During a systems check, Scott reported the front wheels weren't working, and there was no instrument reading on one of the batteries. But because of the redundancy features engineers had developed, the EVA was still GO.

Scott pushed the T-bar upward, and off he and Irwin went on the first wheeled journey on the moon. Three decades later, Morea still struggles to find words to describe his emotions that day: "It was so awesome, it brought tears to my eyes. I knew how much work had gone into that thing."

He also knew how much negative publicity the LRV had received. Boeing had a contract to build four LRVs, but when it was apparent the Apollo program was going to end with Apollo 17, plans for the fourth were shelved before it was completed. Total cost: $38.6 million. The original incentive-based contract was for $19 million. "Even [television talk show host] Mike Douglas had a special on the LRV, talking about how much it was going to cost," Morea says. "Congress was upset about Boeing going over the original budget, so I had the threat of a Congressional investigation hanging over my head the whole time it was being built."

Apollo 15 astronaut Jim Irwin stands beside the Lunar Rover as the lunar mountains loom in the background. Apollo 15 was the first time astronauts used a "car" on the moon. It enabled astronauts to explore larger areas. [Courtesy NASA]

But Morea didn't have much time to reflect. He wanted to know why the front wheels weren't operable.

"It was a restless night for me, I assure you," he says. "We looked at everything and never really found out why. But I have my own theory. I think that as [Scott and Irwin] were going through their checklists and pushing switches, they got one in the wrong position. Because when they came back out for the second EVA after a night's sleep, they went through the checklist again, hit all the switches correctly, and the front wheel steering was working. No problem at all.

"Dave Scott even called back during the second EVA and said, 'I'm amazed the front steering is working. I'll bet some of you Marshall guys came up here and fixed it while we were sleeping.'"

The LRV worked perfectly the rest of the mission, carrying the astronauts at one point 2.1 miles from their lunar module. Scott and Irwin used the car to drive up to the rim of Hadley Rille, a canyon that runs seventy miles across the moon and is three-quarters of a mile wide and 1,100 feet deep. Morea put it in perspective, using his New York upbringing. "If you put the Empire State building down in it, the top of the building would come up to the crater's rim," Morea says. "I still believe they could've driven down in there a bit, but the safety people wouldn't hear of it."

At the end of the first EVA, Scott said he could easily see the LRV's tracks and was going to follow them back to the lunar module, *Falcon*. Ground controllers intervened and asked him to use the accuracy of the navigation system instead. Both astronauts still had ample supplies of water and oxygen. After leisurely traveling up and down a few slopes at about seven miles per hour, Scott and Irwin saw the sun glistening off *Falcon*, their friendly, four-legged home. Again, Morea and his team celebrated.

Apollo 15, 16, and 17 made excellent use of the LRV, covering nearly forty miles on the two missions and collecting approximately 620 pounds of samples—or 74 percent of the 842 pounds brought back by the twelve American explorers. Astronauts on the final three flights reached sections of the moon via the LRV that otherwise would have been impossible.

Apollo 17 astronaut Jack Schmitt, the only scientist to ever explore the moon, says the LRV "proved to be the reliable, safe, and flexible lunar exploration vehicle we expected it to be. Without it, the major scientific discoveries of Apollo 15, 16, and 17 would not have been possible, and our current understanding of lunar evolution would not have been possible."

Those discoveries included rocks 4.6 billion years old; evidence that the moon's craters hold the secrets to the ages of Mercury, Venus, and Mars; and that the moon and earth are formed from different proportions of a common collection of materials.

Scientists continue to study the Apollo collection. Most of it is stored in dry nitrogen at the Johnson Space Center in Houston. A few rocks are locked away in a vault at Brooks Air Force Base in San Antonio.

If Morea has one regret, it is that the LRV's were never given code names like the command and lunar modules. The three cars are simply lumped together and identified only by the mission they were used on. To Morea, it's almost as if three of his children weren't given names at birth. "I guess time was such an issue, that it never came up," he says.

Morea now lives a quiet life in Huntsville, Alabama, with his wife Angela. He will always be recognized among space historians as the man who developed the moon car, and that is a badge he wears with honor and pride. But few realize his other contributions to the space program, particularly his roles in the development of the Saturn V.

At the local health club where Morea works out, people who weren't even alive during Apollo know he used to work at Marshall during the moon missions. They affectionately refer to him as "Rocket Man."

They have no idea how well the nickname fits.

Appendix

U.S. Manned Mission Summary

MERCURY

FREEDOM 7
May 5, 1961
Alan B. Shepard, Jr.
15 minutes, 28 seconds

Suborbital flight that successfully put the first American into space.

LIBERTY BELL 7
July 21, 1961
Virgil I. "Gus" Grissom
15 minutes, 37 seconds

Also suborbital; successful flight, but the spacecraft sank shortly after splashdown.

FRIENDSHIP 7
February 20, 1962
John H. Glenn, Jr.
4 hours, 55 minutes, 23 seconds

Three-orbit flight that placed the first American into orbit.

AURORA 7
May 24, 1962
M. Scott Carpenter
4 hours, 56 minutes, 5 seconds

Confirmed the success of Mercury-Atlas 6 by duplicating flight.

SIGMA 7
October 3, 1962
Walter M. Schirra, Jr.
9 hours, 13 minutes, 11 seconds

Six-orbit engineering test flight.

FAITH 7

May 15–16, 1963
L. GORDON COOPER, JR.
34 hours, 19 minutes, 49 seconds

Completed twenty-two orbits to evaluate effects of one day in space.

GEMINI

GEMINI 3

March 23, 1965
VIRGIL I. "GUS" GRISSOM,
JOHN W. YOUNG
4 hours, 52 minutes, 31 seconds

First manned Gemini flight, three orbits.

GEMINI 4

June 3–7, 1965
JAMES A. MCDIVITT,
EDWARD H. WHITE II
4 days, 1 hour, 56 minutes,
12 seconds

Included first extravehicular activity (EVA) by an American; White's spacewalk was a twenty-two-minute EVA exercise.

GEMINI 5

August 21–29, 1965
L. GORDON COOPER,
CHARLES "PETE" CONRAD, JR.
7 days, 22 hours, 55 minutes,
14 seconds

First use of fuel cells for electrical power; evaluated guidance and navigation system for future rendezvous missions. Completed 120 orbits.

GEMINI 7

December 4–18, 1965
FRANK A. BORMAN,
JAMES A. LOVELL, JR.
13 days, 18 hours, 35 minutes,
1 seconds

When the Gemini VI mission was scrubbed because its Agena target for rendezvous and docking failed, Gemini VII was used for the rendezvous instead. Primary objective was to determine whether humans could live in space for fourteen days.

GEMINI 6-A

December 15–16, 1965
WALTER M. SCHIRRA, JR.,
THOMAS P. STAFFORD
1 day, 1 hour, 51 minutes,
24 seconds

First space rendezvous accomplished with Gemini VII, station-keeping for over five hours at distances from 0.3 to 90 m (1 to 295 ft).

GEMINI 8

March 16, 1966
NEIL A. ARMSTRONG,
DAVID R. SCOTT
10 hours, 41 minutes, 26 seconds

Accomplished first docking with another space vehicle, an unmanned Agena stage. A malfunction caused uncontrollable spinning of the craft; the crew undocked and effected the first emergency landing of a manned U.S. space mission.

GEMINI 9-A

June 3–6, 1966
THOMAS P. STAFFORD,
EUGENE CERNAN
3 days, 21 hours

Rescheduled from May to rendezvous and dock with augmented target docking adapter (ATDA) after original Agena target-vehicle failed to orbit. ATDA shroud did not completely separate, making docking impossible. Three different types of rendezvous, two hours of EVA, and forty-four orbits were completed.

GEMINI 10

July 18–21, 1966
JOHN W. YOUNG,
MICHAEL COLLINS
2 days, 22 hours, 46 minutes
39 seconds

First use of Agena target-vehicle's propulsion systems. Spacecraft also rendezvoused with Gemini VIII target-vehicle. Collins had forty-nine minutes of EVA standing in the hatch and thirty-nine minutes of EVA to retrieve experiment from Agena stage. Forty-three orbits completed.

GEMINI 11

September 12–15, 1966
CHARLES "PETE" CONRAD, JR.,
RICHARD F. "DICK" GORDON
2 days, 23 hours, 17 minutes,
8 seconds

Gemini record altitude, 1,189.3 km (739.2 mi), reached using Agena propulsion system after first orbit rendezvous and docking. Gordon made thirty-three-minute EVA and two-hour standup EVA. Forty-four orbits.

GEMINI 12

November 11–15, 1966
JAMES A. LOVELL, JR.,
EDWIN E. "BUZZ" ALDRIN, JR.
3 days, 22 hours, 34 minutes,
31 seconds

Final Gemini flight. Rendezvoused and docked with its target Agena and kept station with it during EVA. Aldrin set an EVA record of five hours, thirty minutes for one spacewalk and two stand-up exercises.

APOLLO

APOLLO 7

October 11–22, 1967
WALTER M. SCHIRRA, JR.,

Orbited the earth 163 times and performed a thorough check of the command module's systems.

Donn F. Eisele, R. Walt Cunningham
10 days, 20 hours, 9 minutes

APOLLO 8
December 21–27, 1968
Frank A. Borman, James A. Lovell, Jr., Williams A. Anders
6 days, 3 hours, 1 minute

First manned flight around the moon. Made ten orbits of the moon on Christmas Eve. Took photographs to help plot landing sites for future missions.

APOLLO 9
March 3–13, 1969
James A. McDivitt, David R. Scott, Russell L. Schweickart
10 days, 1 hour, 1 minute

Earth-orbit test of the lunar module. First rendezvous between the command module and lunar module. Schweickart performed a thirty-eight-minute EVA to test the lunar spacesuit and backpack.

APOLLO 10
May 18–26, 1969
Thomas P. Stafford, John W. Young, Eugene A. Cernan
8 days, 0 hours, 3 minutes

While performing a dress rehearsal for a lunar landing, Stafford and Cernan flew the lunar module within nine miles of the moon. Young became the first person to solo around the moon.

APOLLO 11
July 16–24, 1969
Neil A. Armstrong, Michael Collins, Edwin E. "Buzz" Aldrin, Jr.
8 days, 3 hours, 18 minutes

First lunar landing. Armstrong became first human to walk on the moon. He spent two hours, thirty-one minutes outside the lunar module; Aldrin about two hours. Collected 47.7 pounds of samples.

APOLLO 12
November 14–24, 1969
Charles "Pete" Conrad, Jr., Richard F. "Dick" Gordon, Alan L. Bean
10 days, 4 hours, 36 minutes

Survived two lightning strikes during launch. Second lunar landing. Touched down, as planned, approximately 600 feet from the unmanned Surveyor 3 explorer. Conrad and Bean performed two EVAs of seven hours, forty-five minutes. Collected 75.7 pounds of samples.

APOLLO 13
April 11–17, 1970
James A. Lovell, Jr.,

Mission aborted when an oxygen tank exploded inside the command service module

JOHN L. "JACK" SWIGERT, JR.,
FRED W. HAISE, JR.
5 days, 22 minutes, 54 minutes

while on the way to the moon. Made a free return around the moon.

APOLLO 14

January 31–February 9, 1971
ALAN B. SHEPARD,
STUART A. ROOSA,
EDGAR D. MITCHELL
9 days, 0 hours, 2 minutes

Third lunar landing. Shepard and Mitchell performed two EVAs of nine hours, twenty-one minutes. Shepard became the only human to ever hit a golf ball on the moon. Collected 94.4 pounds of samples.

APOLLO 15

July 26–August 7, 1971
DAVID R. SCOTT,
ALFRED W. WORDEN,
JAMES B. IRWIN
9 days, 0 hours, 2 minutes

Fourth lunar landing. First to use the Lunar Roving Vehicle, a battery-powered car. Scott and Irwin performed three EVAs totaling eighteen hours, thirty-three minutes. Collected 169 pounds of samples.

APOLLO 16

April 16–27, 1972
JOHN W. YOUNG,
T. KENNETH MATTINGLY II,
CHARLES M. DUKE
11 days, 1 hour, 51 minutes

Fifth lunar landing. First exploration of the lunar central highlands. Young and Duke performed three EVAs totaling twenty hours, fourteen minutes. Collected 208.3 pounds of samples.

APOLLO 17

December 7–19, 1972
EUGENE A. CERNAN,
RONALD E. EVANS,
HARRISON H. "JACK" SCHMITT
12 days, 13 hours, 51 minutes

Sixth and final lunar landing. First U.S. night launch. Cernan and Schmitt, the first true scientist to walk on the moon, performed three EVAs totaling twenty-two hours, two minutes. Collected 243.1 pounds of samples.

Glossary

Apollo-Soyuz—First manned spaceflight conducted jointly by the United States and the Soviet Union during the era of detente. On July 17, 1975, the U.S. Apollo spacecraft and Soviet Soyuz vessel docked for two days of joint operations.

biological isolation garment (BIG)—A special suit worn during recovery by the first three Apollo crews to return from a lunar landing. It isolated the astronauts from earth atmosphere in case they had come in contact with lunar pathogenic bacteria.

command module—The Apollo spacecraft consisting of a working and living area for three astronauts. It is the only part of the entire vehicle that returns from space.

command service module—The compartment attached to the rear of the command module that houses the spacecraft's electrical power subsystem, reaction control engines, part of the environmental control subsystem, and main propulsion engine.

EVA (extravehicular activity)—Anything done outside the confines of the spacecraft by an astronaut. This includes moonwalks and working in open space.

FOPs—Flight Operations Panel meetings in which mission rules were discussed and adopted.

Gemini—The two-men space mission program preceding Apollo used to learn the intricacies of EVAs, rendezvous, and docking. It also helped determine the effects zero gravity had on the human body.

Kennedy Space Center—NASA's manned launch site.

LRV (Lunar Roving Vehicle)—battery-powered moon cars used on the final three Apollo missions to allow the astronauts to explore larger areas of the lunar surface.

lunar module (LM)—The spacecraft specifically used for the moon landing. It included two engines, one for landing and another for blasting off from the moon.

Manned Spaceflight Center—Now known as the Johnson Space Center, it is located in Houston and is the headquarters for all manned missions.

Mercury—First U.S. manned space program involving one astronaut per flight.

mission control—The central communications site for all manned flights.

Mobile Quarantine Facility (MQF)—the trailer-like facility where the first three Apollo crews to land on the moon were detained until doctors determined they were not contaminated with lunar bacteria.

Skylab—The United States space station occupied in earth orbit by astronauts during 1973 and 1974. Skylab was destroyed during reentry into earth's atmosphere in 1977.

Transearth Injection (TEI)—A burn that put the spacecraft on a trajectory from the moon to the earth.

Translunar Injection (TLI)—A burn that put the spacecraft on a trajectory from the earth to the moon.

VAB (Vehicle Assembly Building)—Located at Kennedy Space Center, this massive building is where the Saturn launch vehicle was assembled in stages.

Bibliography

British Broadcasting Corporation, UK. *The Saturn/Apollo Stack*. Updated 2004, http://www.bbc.co.uk/dna/h2g2/A429040.

Chaikin, Andrew. *A Man on the Moon*. New York: Penguin Books, 1994.

Cortright, Edgar M., ed. *Apollo Expeditions to the Moon*. Washington, D.C.: NASA, 1975.

Godwin, Robert, ed. *Apollo 11: The NASA Mission Reports (Volume One)*. Burlington, Ontario: Apogee Books, 1999. http://www.aerospaceweb.org/.

Lindsay, Hamish. *Tracking Apollo to the Moon*. New York: Springer, 2001.

Mailer, Norman. "A Fire on the Moon." *Life*, August 29, 1969, pp. 25–41.

National Aeronautics and Space Administration. *Apollo 11 35th Anniversary*. Washington, D.C.: NASA, 2004.

———. *Apollo Lunar Surface Journals*. Updated 2005, http://www.hq.nasa.gov/office/pao/History/alsj/frame.html.

———. *The Apollo Program 1963–1972*. Updated 2005, http://nssdc.gsfc.nasa.gov/planetary/lunar/apollo.html.

———. *NASA—Exploration Systems*. Updated 2005, http://www.exploration.nasa.gov/articles/.

———. *NASA History Series (SP-4000)*. Created in 2000, http://history.nasa.gov/SP-4029/contents.htm.

———, Media Resource Center, Johnson Space Center in Houston. *Voice Transmission Recordings of* Apollo 8 *and* Apollo 11 *Flights*.

Index

About the Author

BILLY WATKINS, a lifelong Mississippian, has been a newspaper reporter for three decades in his home state, telling the stories of its people. After earning a journalism degree from the University of Mississippi, he was a sportswriter from 1975 to 1990 at *The Meridian Star*, *Jackson Daily News*, and the *Jackson Clarion-Ledger*, and was voted by his peers the state's Sportswriter of the Year three times. He then moved to general features at the *Clarion-Ledger*, where his work has earned him more than forty regional and national awards. He proudly reminds people that the Saturn V rockets, which powered our astronauts on their way to the moon, were all tested in Mississippi, at Stennis Space Center.